4320   2984

W9-BRX-926

WITHDRAWN

Powering the Future

Powering the Future

*A Scientist's Guide to
Energy Independence*

Daniel B. Botkin
with Diana Perez

Vice President, Publisher: Tim Moore
Associate Publisher and Director of Marketing: Amy Neidlinger
Editorial Assistant: Pamela Boland
Acquisitions Editor: Kirk Jensen
Operations Manager: Gina Kanouse
Senior Marketing Manager: Julie Phifer
Publicity Manager: Laura Czaja
Assistant Marketing Manager: Megan Colvin
Cover Designer: Chuti Prasertsith
Managing Editor: Kristy Hart
Project Editor: Lori Lyons
Copy Editor: Apostrophe Editing Services
Proofreader:
Indexer: Erika Millen
Compositor: Nonie Ratcliff
Manufacturing Buyer: Dan Uhrig
© 2010 by Daniel B. Botkin
Pearson Education, Inc.
Publishing as FT Press
Upper Saddle River, New Jersey 07458

FT Press offers excellent discounts on this book when ordered in quantity for bulk purchases or special sales. For more information, please contact U.S. Corporate and Government Sales, 1-800-382-3419, corpsales@pearsontechgroup.com. For sales outside the U.S., please contact International Sales at international@pearson.com.

Company and product names mentioned herein are the trademarks or registered trademarks of their respective owners.

Printed in the United States of America

First Printing March 2010

ISBN-10: 0-13-704976-5
ISBN-13: 978-0-13-704976-9

Pearson Education LTD.
Pearson Education Australia PTY, Limited.
Pearson Education Singapore, Pte. Ltd.
Pearson Education North Asia, Ltd.
Pearson Education Canada, Ltd.
Pearson Educatión de Mexico, S.A. de C.V.
Pearson Education—Japan
Pearson Education Malaysia, Pte. Ltd.

Library of Congress Cataloging-in-Publication Data

Botkin, Daniel B.
  Powering the future : a scientist's guide to energy independence / Daniel Botkin.
      p. cm.
  Includes bibliographical references and index.
  ISBN 978-0-13-704976-9 (hardback : alk. paper) 1. Power resources—United States.
2. Power resources—United States—Forecasting. 3. Renewable energy sources—
United States. 4. Energy policy—United States. I. Title.
  TJ163.25.U6B68 2010
  333.790973—dc22
                                    2009052805

We have only two modes—complacency and panic.
—James R. Schlesinger, the first U.S. Energy Secretary,
commenting on the country's approach to energy (1977)

Contents

Acknowledgments

I wish to acknowledge the important contributions of the following and express my thanks and appreciation for their contributions. First, my wife, Diana Perez Botkin, worked closely with me on the manuscript. She is a professional editor with extensive experience, and we both saw this book as having a potentially useful role and were devoted to it. The extent to which the book is readily readable and accessible is due to her, and also to my sister, Dorothy B. Rosenthal, retired professor of science education, Long Beach State University, who read much of the material more than once. Pat Holl provided invaluable help in obtaining permissions for the book's illustrations.

Matthew Sobel, William E. Umstattd Professor, Chair of Department of Operations, Case Western Reserve University, provided insightful and important comments and suggestions about the economic analyses and cost forecasting, and carried out some essential economic calculations used in this book.

Kenneth P. Green, resident scholar, American Enterprise Institute, checked my calculations to make sure they were accurate.

My son, Jonathan D. Botkin, checked many calculations regarding alternative energy and helped with the interpretation of technical information. Robert Elliott, former Deputy Secretary of State, State of New York, and past mayor of Croton-on-Hudson, N.Y., helped me understand how best to communicate the ideas in this book to those involved in government and politics.

Professor Brian Skinner, Yale University, Department of Geology, guided me to experts on mineral resources. Dr. Franz J. Dahlkamp, author of *Uranium Deposits of the World* (originally published in 1993, new edition published February 2009) and member of the Advisory Board of Strathmore Minerals Corporation, corresponded with me about uranium resources, as did James K. Otton, Ph.D., Uranium Resource Specialist, U.S. Geological Survey. Tom Payne, director of the Ferroequus Railway Company Limited and president of the Great Northern Pacific Railway, Inc., provided essential information about costs of railroad construction. Professor Kingsley Haynes, Dean, School of Public Policy, George Mason University, helped with data about automobile transportation. Historian Dr. Alfred Runte, an authority on the history of railroads, helped me contact railroad experts. Ted Scudder, an anthropologist from Caltech and author of *The Future of Large Dams,* helped with my analysis of major new hydropower installations and the potential or lack of potential for more.

The following read all or parts of the manuscript and provided invaluable suggestions: Kenneth L. Purdy, Charles Sansone, Dr. Bruce Hector, and John Gibson; and Nancy E. Botkin. Dr. John H. DeYoung Jr., chief scientist, Minerals Information Team, U.S. Geological Survey,[1] helped to find reliable information about uranium and thorium ores and the economics and future of nuclear energy. Tom Veblen, Cargill Corp. executive (retired) and head of Superior Business Firm Roundtable, author of *The Way of Business,* helped me understand the perspective of a major corporate business executive on energy and society.

Of course, all statements, calculations, and conclusions are my own and I am responsible for any errors, but I have done my best to avoid any errors in calculations. No one and no organization, foundation, or any other entity has supported this book except the publisher. It is therefore a totally independent assessment by the author.

About the Author

Daniel B. Botkin is Professor (Emeritus), Department of Ecology, Evolution and Marine Biology, University of California, Santa Barbara, and President, The Center for The Study of The Environment, a non-profit corporation that provides independent, science-based analyses of complex environmental issues. The *New York Times* has called him "one of the world's leading environmental researchers," who has "done much to popularize the concept of using yet maintaining the world's natural resources."

His research includes creating the first successful computer simulation in ecology; studies of wilderness and natural parks ecosystems—from the Serengeti Plains of Africa to the Boundary Waters Canoe Area of Minnesota and Isle Royale National Park; threatened and endangered species—whooping cranes, salmon, bowhead and sperm whales, and African elephants. He was among the first to investigate possible ecological effects of global warming and to help NASA use satellite imaging to study the Earth's global environment.

His dozen books include *Discordant Harmonies: A New Ecology for the 21st Century*, which "is considered by many ecologists to be the classic text of the [environmental] movement," according to *The New York Times*. *Beyond the Stony Mountains* describes nature in the American West before European settlement, based on the journals of Lewis and Clark. *No Man's Garden* analyzes the value of nature and the relationship between people and nature. He has published op-ed pieces in many major newspapers concerning global warming, biological diversity, and energy, and more than 150 scientific papers.

His recent awards include the Astor Annual Lectureship for 2007 (Oxford University); annual distinguished visiting scholar for 2008 (Green Mountain College, Vermont); and the Long Beach Aquarium, Long Beach, California, has appointed him its first-ever distinguished visiting scientist for November 2008. He is also the recipient of the Fernow Award for Outstanding Contributions in International Forestry and the winner of the Mitchell International Prize for Sustainable Development.

His other academic appointments include Professor of Biology and Director of The Program in Global Change at George Mason University, Fairfax, VA; Professor of Systems Ecology at the Yale School of Forestry and Environmental Studies; and Research Scientist, The Ecosystems Center, Woods Hole, MA. His degrees are B.A. (Physics; University of Rochester), M.A. (Literature, University of Wisconsin), and Ph.D. (Biology, Rutgers University).

Preface

What this book is

This book is about how to solve our energy problem. It presents the facts concerning our energy needs, desires, and supplies, and the environmental and human effects of obtaining and using energy. It also includes calculations and analyses based on these facts. The purpose of the book is to provide U.S. citizens and our elected representatives with information that will enable us to make rational, economically and environmentally sound decisions.

What is our energy problem, and why do we have it? Some people believe that energy is a problem only because continued burning of fossil fuels will lead to undesirable global warming. But even without that possibility, the people of the world, and especially the people of the United States who use more energy than any other nation, have an energy problem: The need and desire for energy will increase faster than it can be provided from standard sources. As the world's population grows, and as living standards and expectations rise, people around the world will want more energy. Meanwhile, petroleum is limited and, according to estimates by petroleum geologists and economists, is likely to become so rare by 2050 that it may be too expensive for most energy applications. Pollution from fossil fuels will also continue to be a problem and is likely to worsen with increased mining and use. As the human population continues to grow and as quality-of-life expectations rise, competition for land and water will also increase. For the United States, military and economic security are also strong reasons for seeking energy independence, or as close to that as possible.

Some believe that we should become energy minimalists, each person using as little energy as possible. This book takes a different tack. It assumes that we should use energy as efficiently as possible, but that abundant energy is necessary for the quality of life that people today expect. Minimal energy would be that required to provide food, water, shelter, and access to medical care. Civilizations require more. People need to be educated; funds are needed for arts, humanities, sciences, and for recreation and entertainment. Life should be joyful, and music, dance, and the graphic arts don't come energy-free. If people have such limited energy available that they can only focus on the bare necessities of life, they don't have time to think about who to vote for, how to organize and run a political campaign, go to town meetings, and so on—the things that we take for granted but are necessary for a democracy. Thus it also may be that democracies benefit from, or even require, more than the minimal energy required for the barest human survival.

Of course, there is a wide range between just enough energy to get food, water, and shelter, and to have everything anyone could ever want—and the decision about what is a quality of life and what is extravagance is a question that goes far beyond this book. My point only is that we cannot have peace, culture, joy, and civilization, let alone freedom and democracy, without energy that allows us to do more than just survive. In the last chapter, I consider various scenarios, some of which contrast very great differences in per capita energy use.

And abundant and superabundant energy can be used for evil as well as good, fueling wars, vast armies, oppressive dictatorships, and terrorism, the worst of man's actions. But without enough energy, no one can work for what is good about and for people. Energy is the key. Physicists define energy as the capacity or the ability to move matter, which means it is necessary for a human being to do anything, including those things that are worthwhile and good.

The good news is that our energy problem can be solved: Today's technology can solve it. America can be energy-independent—our nation can provide a sufficient and sustainable supply of energy with relatively little change in the quality of our lives or in our overall standard of living. Indeed, done with great care, it will result in a better environment and an improved quality of life for most of us.

The tough news is that achieving this energy independence will be expensive and will require a national commitment that is unusual for a democracy, although not unprecedented. Examples of such commitment in our past include the expansion of European-based civilization westward with the Louisiana Purchase, the response of the nation to the Great Depression of the 1930s and to World War II, and our success in putting a man on the moon ten years after we decided to do so. However, such concerted national efforts are rare. Mostly, we tend to muddle through, often waiting to rebuild a bridge until the old one finally collapses.

The solution requires informed citizens and informed politicians. It requires political will and individual personal commitments in ways that we have not chosen to seek in past decades. A necessary and important part of energy independence is clear thinking, rational, science-based analysis. It requires innovation, creativity, invention, and entrepreneurship. This book is meant to be a foundation for the path to energy independence.

What this book is not

Two energy-related topics that are mentioned but not discussed in depth are carbon offsets and subsidies. The subject of subsidies is so big and complicated that a separate book would be needed to analyze the subsidies for each source of energy from cradle to grave—from discovery, invention, and exploration, to use and the problems of dealing with the wastes and land conversions. The same holds true for carbon offsets. By seeking a way to replace all fossil fuels, this book does point the way to reducing the production of carbon dioxide from burning fossil fuels. It just does not deal with the particulars and differences in carbon dioxide releases among different energy sources. There is no intention to minimize the importance of either of these topics.

The focus here is on how energy supply can be obtained technologically in a way that is as environmentally beneficial and benign as possible. Although I do discuss costs to some extent, this is not a book primarily about economics. My hope is that this book will provide a basis from which economists and others can consider in much greater detail the economic consequences of difference choices and the ways that a society and individuals can be motivated to choose and work toward a specific solution.

The energy solution has become a major political and ideological debate, and obviously a great deal of money, influence, and direction of our society is at stake. During the past few years, I have been asked to discuss the solution to energy supply in a number of forums, sometimes as part of a panel discussion or debate. It should not be surprising that many approach this important topic from a specific political and ideological goal, which leads them to pick the facts that support these goals and ignore those that don't. I believe that this approach is a road to failure. Of course, no one can be completely free of prejudgments and emotional assumptions, but to solve a large-scale technological problem like energy supply requires a rational approach. We have to be careful to see this as more of an engineering problem, solidly based in science, than an expression of a political philosophy or ideological conviction.

Why I wrote this book

As an ecologist with a background in physics, and as chairman of the Environmental Studies program at the University of California, Santa Barbara, I have long been interested in how energy is obtained and used in natural ecosystems, how energy from our environment affects us, and how we affect our environment in our pursuit of energy. For my work, I had to keep up with energy issues, and in doing so noticed some odd contradictions that began to occur around 2002. Solar and wind were already providing energy in many parts of the world, but environmental economists I worked with kept telling me a different story. "The conventional wisdom," they said, was that solar and wind power can never amount to anything.

At that time, my son, Jonathan, worked for a company called PowerLight, which manufactured and installed some of the largest solar energy facilities in the world. (He had previously worked for U.S. Windpower.) Deciding it might be interesting for all involved, I set up a series of telephone conferences between the environmental economists and the PowerLight engineers. Each time one of the economists asked a question about solar power and was answered by an engineer, the economist would reply, "But according to conventional wisdom, solar and wind energy can never amount to anything." This went on for three weeks of conference calls, until finally the engineers and I gave up, discouraged because nobody seemed interested in the facts.

Soon afterward, the *New York Times* published an interview with James Lovelock, the famous British chemist and environmentalist who came up with what he called "the Gaia Hypothesis," an expression of the idea that we are all connected to all of life by a planetary system. Commenting on the energy problem, Lovelock said, "If it makes people feel good to shove up a windmill or put a solar panel on their roof, great, do it. It'll help a little bit, but it's no answer at all to the problem."

Meanwhile, hundreds of thousands of people in developing nations were buying or building cheap solar and wind devices to provide them with enough electricity to cook their food, run computers and some small home appliances, and thereby join the modern age. It wasn't just making them feel good; it was improving the fundamentals of their lives.

With all the debate about our energy supply, the end of the era of cheap and abundant oil and gas, and concerns about global warming, I decided to look at each form of energy: how much is available, how much we now use, how much we will use in the future, and what our options are to move away from fossil fuels. The approach would be the same one I have used for all scientific problems: looking at the most reliable data available and making the obvious calculations and analyses. In the past, I'd always been surprised by what the facts revealed, because so often what they told me contradicted the conventional wisdom. In many cases, the facts—especially quantitative information—necessary to reach a conclusion are completely lacking because nobody ever bothered to get them, and without these, of course, even the simplest calculations and analyses haven't been done.

With so many people talking and writing about the energy issue, why should you pay attention to this book? Because to the best of my ability I have hunted down the most solid facts and information and analyzed them as carefully and as free of my personal biases as possible, searching out the facts, doing the calculations, and checking those measurements and calculations with experts in their fields. You will surely find some of the results surprising, and I hope you will also find them helpful.

Introduction

Blackout!

FIGURE I-1   The big blackout of 2003. A bright full moon over a darkened New York City skyline during the blackout that started Thursday, Aug. 14, 2003, and affected 80,000 square miles in the eastern United States and Canada. *(Source: Bob Gomel/Time & Life Pictures/Getty Images)*

Thursday, August 14, 2003

It was one of those muggy New York City summer days that began like all the others. But it ended in a way that was an eerie precursor of the electric-power problems that would have major impacts on New Orleans after Hurricane Katrina in August 2005, and remain an ominous threat to all of us who live in a modern industrialized society.

Our apartment in Manhattan's Chelsea neighborhood, west of Fifth Avenue between 34th Street and 14th Street, had an unobstructed view of the city from the 20th floor but did not have air conditioning. It had "air cooling"—cool air pumped from a central energy plant through underground steam pipes and up into our apartment—and on this hot day it was doing its usual mediocre job, making it only about 5 degrees cooler inside than out. Indeed, as the morning wore on, our eastward-facing terrace fell into shade and actually became more comfortable than the apartment. We leaned against the railing and looked up at the Empire State Building just to the northeast.

As always, Manhattan was an impressive sight, a seemingly invincible metropolis, a triumph of modern civilization over raw nature. Our use of abundant and cheap energy was everywhere evident: in the streets teeming with taxis, buses, and trucks; in the sounds of big air-conditioning units on rooftops, jackhammers, and street cleaners—sounds of internal combustion engines of every size. Even indoors we could hear our heating/cooling convectors blowing, the elevators running, the water pumps thumping through the walls, an amplified guitar played by an upstairs neighbor, someone's vacuum cleaner, and power tools being used to renovate an apartment somewhere in the building.

But unknown to us, hundreds of miles away in the Midwest something was going wrong that would soon affect this great city and thousands of square miles around it.[1]

Noon: New York City's temperature climbed to 91°F, the hottest so far that month. It hadn't rained to amount to anything for ten days, and it hadn't been this hot since July 5, when the temperature climbed to 92° F. Today, the sun shone through a gray-blue haze.[2]

It wasn't hot just in New York City; it was hot throughout the northeastern United States and adjacent Canada, and air conditioners across these thousands of square miles drew huge amounts of electricity. That electricity flowed over a gigantic grid system, thousands and thousands of miles of high-tension wires across the Midwest and eastern United States and Canada, from the Great Lakes of Michigan to the Atlantic Ocean shores of Canada. The Midwest Independent Transmission System managed the huge electrical grid of the eastern United States and Canada for about 30 big power companies.

In elementary school, we were given a pamphlet from the local power company that showed a picture-book idealization of how electricity gets to your house. There was a generating station, usually shown as a hydroelectric-power dam, with wires coming out of it that ran along tall high-tension poles across the countryside to your town and then through a series of transfer points to a telephone pole outside your house and then to your home. It seemed simple: a place that made power, a way to transmit it, and us to use it.

But that isn't how it works anymore for most of North America. Instead, energy from many generating stations flows into a central grid, and this grid then spreads like a complex spider web throughout the countryside so that everybody's home is ultimately connected to everybody's source of energy, more or less. And the sources of energy were becoming more varied every year, with wind turbines and solar energy parks beginning to add electricity to the grid. The giant tangle of wires consists of more than 150,000 miles of interlocked power lines connecting plants that generate more than 850,000 megawatts. This amount of energy is hard to imagine, but look at it this way: One megawatt of electricity is about enough to power 300 homes, so the grid carries enough electricity for 255 million homes.

New York State alone was using 28,000 megawatts. That's more than 37 million horsepower, enough to run 370,000 automobiles starting a race at full acceleration at the same time; enough to power 280 million 100-watt light bulbs, about one bulb for every person in the United States at that time.[3] But unlike those 370,000 cars, all these electrical devices were connected, like the colored lights on a Christmas tree.[4] The surges were huge, too, 3 billion watts surging up and down New York State's high-tension power lines.

The grid has 130 control centers operating 24/7.[5] Most of the time everything works, but on this day at around noon, hundreds of miles away from New York City in Carmel, Indiana, Don Hunter, one of the coordinators of the Midwest Independent Transmission System, saw that there was too much electricity flowing on the wires. If the power load got too high, the grid could break down, even catch fire, just like an overloaded electric circuit in your home.

Concerned, Hunter put in a call to the Allegheny Company, one of the 30 cooperating power producers and distributors his firm coordinated, and asked them to reduce their electrical load on the grid.

Allegheny's representative at first agreed, but then said, "Don, question for you. I got a call from the people at our marketing end. They want to bring on another unit at Wheatland." [6]

"We would have to say no to that, at this point," Hunter responded.

But Allegheny Power went ahead and upped its power production anyway.

1:00 pm: The grid started to unravel. In Cincinnati, Ohio, an employee at Cinergy Corporation, another of the on-the-grid electric power producers, called the Carmel, Indiana, coordinating offices.

"Hey, we've got big problems," the Cinergy employee, Spencer, said.[7]

"We don't want no big problems," a center employee responded.

"No, we've got a huge problem," Spencer said, explaining that a major transmission line across Cinergy's system in Indiana had gone down, and the power moving east through the state was endangering other lines. To protect the still functioning lines, Cinergy wanted generators in the western part of the state to cut their electrical output and asked that generators to the east simultaneously increase production, assuring that enough power would continue to be available but taking some of the load off the remaining power lines. "We need to get something under control here....We're setting for bigger problems if we don't get this under control quick," Spencer said.

4:10 pm: The entire northeastern power grid started to become unstable. Electric power flow suddenly reversed direction between Michigan and Ontario, and then power started to oscillate all over the grid, the flow increasing and decreasing rapidly, tripping safety circuit breakers. New England and the Canadian Maritime circuit switched off the main grid immediately, and this saved them, so they kept working independently. But the oscillations set off a cascading blackout throughout the rest of the grid, starting somewhere in the Midwest and shutting down electricity in Ohio and Michigan and then on to other states. Suddenly, power went out in eight states and the Canadian province of Ontario, creating North America's largest blackout.

Back on our terrace in New York City, we saw traffic lights on Eighth Avenue wink out. Big floodlights on skyscrapers went dark. Although we didn't know it, more than 100 power plants had just shut down—80 fossil-fuel plants and 22 nuclear. In New York City, 6.7 million customers lost electricity in a few minutes. In the Northeast, 50 million people lost electric power.

4:17 pm: The loss of air conditioning on a hot summer day would have been bad enough, but much more than that was gone. Take Detroit, for example, right in the center of the blackout, and where events were well recorded. The blackout hit Detroit at 4:17 p.m. One thing about electricity, it moves fast, and when it goes, it goes quickly.[8] The city's airport shut down because all its lights and electronics were powered by electricity from the grid. Northwest Airlines alone would soon have to cancel 216 flights in and out of Detroit Metropolitan.

Rush-hour commuters were stalled everywhere. Perhaps the worst spot to be driving was in or into the Detroit-Windsor Tunnel between Michigan and Ontario. About 27,000 commuters used it daily. Some were stuck in the dark. People waited seven hours in the line to go through.

Amtrak trains stopped running; the railroad was without electric signals, and, even more surprising, no one had any idea where any train was. Even the main train from Detroit to Chicago was lost temporarily. You might ask why people stuck on trains didn't use their cell phones to give their trains' locations. The answer is that all cell phones stopped working too—the towers that sent and received their signals were powered by the grid. Even Detroit's homeland security director couldn't use his cell phone. The city was suddenly much more vulnerable to terrorism.

Water became a problem. Half of the Detroit region's residents, 4 million people, were suddenly without water because water in Detroit is moved by electric pumps also running off the grid. An arsonist set fire to a two-story duplex, but without water pressure the city's firemen could do little. Worse, an explosion occurred because of the blackout at the Marathon Ashland Refinery *in* the city. At the moment, the firemen couldn't do anything about it.

Although hospitals have backup generators, in some cases they weren't enough. A backup generator at the North Oakland Medical Center broke down and sent smoke through a hospital in Pontiac, Michigan, and 100 patients had to be evacuated.[9] Fortunately, the hospital's emergency vehicles had fuel even though all the gas stations had stopped pumping.

You couldn't buy gasoline, because at that time gas stations had only electric pumps connected to the grid. (Early in the 20th century, gas stations used pumps worked by hand.)

Dusk, August 14: The view from our lofty perch showed an island of light—Penn South, our ten-building cooperative complex—in a sea of darkness that was the rest of the city. Even the colored lights atop the Empire State Building had gone dark. Our island of light existed because Penn South has its own electrical generating station that operates both on and off the grid and thus was running despite the blackout. It hadn't yet occurred to me that such an off-the-grid electrical generator might be part of a future solution to our nation's growing energy needs.

Friday, August 15: The power outage continued the next day. In Detroit, the more imaginative did a little creative thinking and made some money. Tim and Deb McGee opened a breakfast bar in their drive-way with a row of tables and chairs.[10] Others did the usual price gouging. According to the *Detroit Free Press,* a skinny kid stood on a street corner in Dearborn, Michigan, waving a water bottle and shouting, *"Wattaa wattaa, wattaaaaaaa! Only two bucks!"* Reading about this, I thought of the New Hampshire house I'd once lived in that had an old-fashioned water pump in the kitchen. Even back in 1963 friends thought we must be a little crazy to depend on that antique, but folks in Detroit certainly could have used a few of them on August 15, 2003. At the least, it raised the question of whether there weren't alternatives to a single grid, per-haps a mix of energy sources that would give our cities and our nation better energy reliability and security.

Most amazing is how quickly so much that we take for granted about modern civilization and its technology was suddenly revealed to be fragile.

This event, our nation's largest blackout, was short-lived, but it demonstrated two truths we have been reluctant to face: how dependent we are on cheap, easily available energy and how vulnerable our one huge, complex, interconnected energy-supply system is.

By Friday afternoon, a lot of the power had been restored in Michi-gan. By 9:00 p.m., power was restored to New York City and adjacent Westchester County to the north. It hadn't been a long blackout, but it had covered a large area and, despite its brevity, had many effects.

Why did the lights go out?

Spencer and others who kept the giant grid running knew the breakdown was not an "accident" in the ordinary sense. It was the disastrous result of a series of events and a set of conditions that were well understood by those in charge.

There are basic problems with the grid as it exists today. First, it's getting old—few of the transmission lines are younger than 15 years.

Second, in recent years it wasn't being cared for; spending to maintain and repair the grid declined. From 1988 to 1997, investment in new transmission lines decreased almost 1% every year, and maintenance spending for existing lines decreased more than 3% per year, while at the same time power demand increased 2.4% a year. With growing interest in improving the grid, some recent developments are encouraging. In 2009, the U. S. Recovery and Reinvestment Act provided $343 million to build a new grid transmission line in the Pacific Northwest to increase the amount of electricity from wind power. But so far these are small advances compared to the overall need.[11]

Third, the grid hasn't kept up with technology. For example, state-of-the-art digital switches, which could respond better and faster in power emergencies, haven't been installed.

Fourth, the grid was built for use in emergencies only—say, when one utility's power plant went down and it needed to temporarily borrow power from another system. But today the grid is used in ways that were not foreseen and for which it was not designed.

Fifth, the grid's control centers cannot force member companies to comply. In the few minutes before the blackout started, employees at the Midwest grid control center were in a bind: Their company was responsible for preventing a collapse, but it couldn't force the member companies to act; it could only try to persuade them. "It would be kind of a voluntary thing," Janice D. Lantz, a spokeswoman for the Hagerstown, Maryland-based Allegheny Company, explained later. Individual companies resisted attempts at centralized control. They wanted local control over their own actions.[12] And they had it.

At the start of the blackout, some key people were not aware of the problem until it was too late. Representatives of the International Transmission Commission said they were unaware of the problem until two minutes before the power went out in Michigan. Detroit Edison made the same claim.

And because there is just one grid in any area, few had any alternative sources of electricity after the grid went down. Here and there families had purchased portable generators powered by small gasoline engines, but most people found these too complicated to use, and many were reluctant to store gasoline in their houses. Throughout most of the nation, there didn't seem to be any alternative. There was the big power system, and if it failed, you suffered. In summer, you turned to your stockpile of candles and bottled water, hoped not to lose everything in your freezer, and simply waited and sweltered. In winter you piled on more clothes, since most home heating systems required electricity to make them run, whether they were fueled by natural gas or oil, the two major fuels to heat America's buildings.

Can we prevent more, and bigger, blackouts?

Plenty of people will tell you that a nation that turns more and more to solar and wind energy is asking for more trouble of this kind. They argue that wind and solar are too variable, that there aren't any good ways to store the electricity they produce, and that massive electrical generation from them will only further destabilize the grid.

People still heating their homes with firewood ask why we don't go back to biological fuels. And there was just the beginning of interest in large-scale farming of crops like corn that would be turned into alcohol to run cars, trucks, and electrical generators. This, they would soon be arguing, was a better way to go because America already had large facilities to store liquid and gas fuels.

Watching the city go dark from our Manhattan apartment, we could see once again how much our modern way of life depends on energy— and not just a minimal amount of energy to help us get food, water, and shelter. We need an abundance of energy for all the aspects of life that people enjoy and depend on, from recreation to health care. I believe we can achieve this.

There are four parts to our energy crisis: (1) lack of adequate sources of energy; (2) the need to move away from dependence on fossil fuels; (3) lack of adequate means to distribute energy safety, reliably, and consistently; and (4) inefficient use of energy, with major environmental effects. We have to solve all four problems and solve them quickly. What do we do first? Can we do it all in time? What is the best energy source? Is there just one that is best, or does the solution lie with some combination of energy sources?

Improved ways of distributing energy are crucial. In the big black-out, gasoline couldn't be pumped at gas stations because few stations had installed small electric generators to run the pumps. The lack of these small generators at gas stations symbolized how our electrical generating system had become centralized. That old technology could have been a backup today, but wasn't. Or a gas station could keep a small gasoline-powered generator, the kind many homeowners have on hand for emergences. Few thought about off-the-grid local energy generation. Most who did were solar- and wind-energy enthusiasts.

A friend who built her house in the hills above Santa Barbara, California, was one of these enthusiasts. Unconnected to the grid, the house stored energy generated from the wind and the sun in a huge array of lead-acid batteries, the same kind that are in your car. But these were housed in large glass cylinders, so you could watch the acidic water bubble as electrical energy flowed into it from the wind turbine and solar cells. These, by the way, generated direct current (DC) electricity. Because most modern appliances run on alternating current (AC), to use anything with an electric motor, such as a vacuum cleaner, she had to have electronics that converted the DC to AC. But that process used a lot of the energy. So her house was wired with two systems, one AC and one DC. The lights ran on the DC.

Because so few houses had taken this off-the-grid route, providing electricity for a house like my friend's was pretty much a do-it-yourself hobby that took a lot of time. I admired it, but at the time it didn't seem likely that much of America could go that way. But during the blackout, a lot of people watching the food spoil in their refrigerators would have been grateful for such a system if they had known about it. Ironically, the great electrical public works projects of the 1930s, meant to bring electricity to the farm as well as the city, pretty much brought an end to local electrical power generation, or any local energy generation other than wood in fireplaces and woodstoves. That's why traveling across America's farmland you see so many of those quaint windmills that used to pump drinking water for cattle now sitting idle, with perhaps a blade or two missing.

At present, about 85% of the total energy used throughout the world, and also in the United States, comes from fossil fuels. Everyone is familiar with the controversies about fossil fuels: Coal, oil, and gas are highly polluting fuels that we use to our detriment as well as our benefit; oil and gas are going to run out, with first oil and then natural gas becoming economically unavailable in the not too distant future; and those concerned about global warming believe we must move away from these fuels. For the United States, which is no longer producing as much oil and gas as it uses, moving away from these fuels is necessary for energy independence, for national security, and for a stable and productive economy.

There are many proponents of each source of energy, each claiming that their favorite source is *the* solution. Petroleum and natural gas enthusiasts say that there is bound to be a lot more of those fuels out there somewhere under the ground and under the ocean. We've always found more in the past. Some say, "Trust us; we will find more."

Biofuels have their enthusiasts as well, ranging from small-farm cooperatives to giant agricorporations. They say, "Trust us; pretty soon the technology will be invented to make our biofuel crops energy-efficient, and we will grow our own energy solution." Solar, wind, and ocean enthusiasts meanwhile ask us to follow them.

Which then is a possible, practical, and reliable solution? In writing this book, I have had to dig out a lot of obscure facts, do a lot of calculations from those, including costs and land area required, and think about what mix of all the sources of energy will be the solution.

First, some terms you need to know

The definition of *energy* seems straightforward enough: Energy is the ability to move matter. So what's so complicated about it? For one thing, it's still a difficult concept. Yes, it's the ability to move matter, but even though we need it and use it all the time, we can't see it—it isn't a "thing" like a table or chair or automobile or computer or cell phone.

For another thing, talking about energy is confusing because of all the terms used to discuss and measure it. It comes in so many different units. At the supermarket, we buy potatoes in pounds or, outside the United States, in kilos; we don't have different measurements for baking potatoes, boiling potatoes, red potatoes, Yukon Golds, and sweet potatoes. Each kind of energy, however, has its own measure. The two teams most familiar to us are calories, when we're trying to lose weight, and watts, when we check what size light bulb we need to buy. But that's just the beginning. Oil is discussed in terms of barrels; natural gas in cubic feet or, worse, in terms of its energy content, expressed in British thermal units (BTUs), an old measure dating back to the beginnings of the Industrial Revolution. Electricity, as well as any energy source used to make electricity, comes in several measures: *watts,* or, more commonly, *kilowatts* (KW), which are thousands of watts; sometimes *megawatts* (MW), millions of watts, to describe the capacity of a generator; and kilowatt-hours or megawatt-hours, the actual energy yield or output. Some people discuss energy in terms of *joules,* a measure of energy originating in Newtonian physics.

And that's not all. People write about huge and unfamiliar numbers, like a quadrillion BTUs (a *quad*) and *exajoules* (don't ask). Even the simple calorie that we're all familiar with, the one listed on food packages, is actually an abbreviation of *kilocalorie.* The real (and little) calorie is the amount of heat energy that raises a gram of water from 15.5°C to 16.5°C. That's so small an amount that dietitians talk in terms of a thousand of these—enough to raise a liter of water (about as much as a medium-size bottle of gin) that same 1°C.

At least we come across calories (that is, *kilo*calories) and watts in our everyday modern life. But we rarely get to compare them. At the fitness center where we go to exercise, the Elliptical Trainer machine does it, showing us the watts and the calories that we generate per minute and that we therefore are using as we exercise. The other day I was using about 130 watts on the machine, enough to run one 100-watt light bulb with a little left over.[13]

A historical perspective

Our current energy crisis may seem unique, but it has happened to people and civilizations before.[14] All life requires energy, and all human societies require energy. Although we can't see and hold energy, it is the ultimate source of wealth, because with enough energy you can do just about anything you want, and without it you can't do anything at all.

Human societies and civilizations have confronted energy problems for thousands of years. Ancient Greek and Roman societies are a good case in point. The climate of ancient Greece, warmed and tempered by the Mediterranean Sea, was comparatively benign, especially in its energy demands on people. Summers were warm but not too hot, winters cool but not very cold. With the rise of the Greek civilization, people heated their homes in the mild winters with charcoal in heaters that were not especially efficient. The charcoal was made from wood, just as it is today. As Greek civilization rose to its heights, energy use increased greatly, both at a per-capita level and for the entire civilization. By the 5th century B.C., deforestation to provide the wood for charcoal was becoming a problem, and fuel shortages began to occur and become common. Early on in ancient Greece, the old and no longer productive trees in olive groves provided much of the firewood, but as standards of living increased and the population grew, demand outstripped this supply. By the 4th century B.C., the city of Athens had banned the use of olive wood for fuel. Previously obtained locally, firewood became an important and valuable import.

Not surprisingly, around that same time, the Greeks began to build houses that faced south and were designed to capture as much solar energy as possible in the winter but to avoid that much sunlight in the summer. Because the winter sun was lower in the sky, houses could be designed to absorb and store the energy from the sun when it was at a lower angle but less so from the sun at a higher angle. Trees and shrubs helped.

The same thing happened later in ancient Rome, but technology had advanced to the point that homes of the wealthy were centrally heated, and each burned about 275 pounds of wood every hour that the heating system ran. At first they used wood from local forests and groves, but soon the Romans, like the Greeks before them, were importing firewood.[15] And again like the Greeks before them, they eventually turned to the sun. By then, once again, the technology was better; they even had glass windows, which, as we all know, makes it warmer inside by stopping the wind and by trapping heat energy through the greenhouse effect. Access to solar energy became a right protected by law; it became illegal to build something that blocked someone else's sunlight.

Some argue today that we should become energy minimalists and energy misers, that it is sinful and an act against nature to use any more than the absolute minimum amount of energy necessary for bare survival. But looking back, it is relatively straightforward to make the case that civilizations rise when energy is abundant and fall when it becomes scarce. It is possible (although on thinner evidence) to argue that in the few times that democracy has flourished in human civilizations, it has done so only when energy was so abundant as to be easily available to most or all citizens.

As a result, in this book I argue for changes in where and how we get and use energy, but I do not argue that we should become energy minimalists or energy misers. On the contrary, I think we need to learn how to use as much energy as we can find in ways that do not destroy our environment, do not deplete our energy sources, and do not make it unlikely that our civilization will continue and flourish in the future.

The path to such a world is possible but not simple, not answered with a slogan, not solved by a cliché. If you value your standard of living and the way of life that our modern civilization provides, with its abundant and cheap energy, follow me through this book as we examine each energy source and the ways in which some can be combined into viable energy systems for the future.

A traveler's guide to this book

Each of the first nine chapters discusses a major source of energy: how much energy it provides today, how much it could provide in the future, how much it would cost, and its advantages and disadvantages.

We begin with conventional fuels—fossil fuels, water power, and nuclear fuels—energy sources that dominated the 20th century. We then go on to the new energy sources, those that may have played small roles in the past but are now viewed as having major energy potential. In addition, we devote a chapter to energy conservation. The last part of the book talks about larger and broader issues that involve, or could involve, some or all energy sources: how to transport energy; how to transport ourselves and our belongings; how to improve energy efficiency in our buildings.

And finally, in the last chapter, I attempt to put the whole thing together in formulating a first approximation of an achievable and lasting solution to our energy problem.

Section I

Conventional energy sources

I use the term *conventional* to mean those energy sources that dominated the 20th century, are familiar to us, and therefore seem conventional. These are the fossil fuels—oil, gas, and coal—as well as water power from rivers and streams, and nuclear power.

In working through the energy issues myself, I kept needing to refer back to the broad view of how much energy each source provided and might provide in the future. Table SI.1 and Figure SI.1 provide a quick summary, which you may want to refer back to. I also found it hard to conceive of energy amounts this large, so I tried to express them as something I could imagine. That's why Table SI.1 shows how many Boeing 747 round-the-world trips each amount of energy could provide.

Table SI.1 U.S. Energy Use 2007[1]

Source	Billions of Kilowatt Hours	Number of Boeing 747 Round-the-World Trips
Coal	6,738	1,458,442
Oil	11,719	2,536,580
Natural Gas	6,738	1,458,442
Nuclear	2,344	507,359
Biofuels	984	212,987
Hydro	861	186,364
Geothermal	103	22,294
Wind	82	17,749
Solar	21	—
Ocean	—	—
TOTAL	**29,590**	**6,400,216**

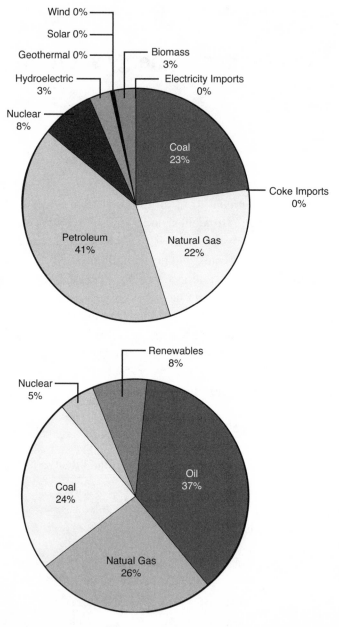

FIGURE SI.1 Total energy consumption in 2007 for (top) in the United States and (bottom) in the world. The zeros for wind, solar, and others in the top chart mean less than 0.5%, not exactly zero.

I have shown the energy use both in a table and in a graph, because some people relate more easily to one and some to the other.

Putting a number on the total amount of energy people use requires the use of two of those terms we previously mentioned. Total energy use in the United States is approximately 29,000 *billion* kilowatt-hours (refer to Figure S.1 and Table S.1). To be specific for our later calculations, the amount we use is 29,590 billion kilowatt hours. Worldwide, people use about 116,000 *billion* kilowatt-hours. Note that the United States uses about 22% of all the energy consumed in the world!

Putting some reality into energy

We talk about large amounts of energy and power, and these have little reality, both because the numbers are big and because energy is invisible, so we can't picture its quantity. Here are a few comparisons that may help. A typical number we will deal with is a million kilowatts of power, which, for example, is the capacity of a typical nuclear power plant reactor. What can we do with that amount of power? If that power plant runs an hour, that gives us 1 million kilowatt-hours, another large and abstract number. Table SI.2 shows what that amount of energy can do.[2] For example, that's the equivalent of more than 9,000 100-watt light bulbs burning around the clock, day in and day out, for every man, woman, and child in the United States.[3] It is also equivalent to more than 9 round-the-world trips by a Boeing 747.

Table SI.2 What You Can Do with the Output from a Typical Large Electrical Power Plant

Activity	Units
Typical electrical power plant output (1,000 megawatts) operating 1 hour	1,000,000 kilowatt-hours
Number of Yankee night games powered by 1-hour's output from that power plant	37 games
Number of round-the-world trips by a Boeing 747 powered by 1-hour's output from that power plant	9.2 trips
Number of hours a 1-megawatt power plant has to run for one Boeing 747 round-the-world trip	4.7 hours
Number of people exercising moderately for 1 hour to provide the same output as that power plant	6,666,667 people
Number of cities the size of Los Angeles to provide that power output if every resident was exercising moderately	1.7 L.A.'s
Number of 100 HP autos to provide the power output of that power plant	13,410 cars

The first simple message from these numbers is: *We use a lot of energy!*[4]

The second simple message is that in industrialized nations, most energy comes from fossil fuels.

Conclusion: Of the conventional fuels, only freshwater provides renewable energy. Nobody doubts that eventually the others—fossil fuels and conventional nuclear fuels—will run out. The important questions follow: When will this happen? What are the environmental and economic effects of these fuels in the near and distant future? How do we make a transition away from them? And when do we have to do that? Economists will tell you that running out of a fuel source isn't simply a question of the total amount in the ground and the rate at which you pull it out. There will always be some amounts of fossil fuels in the ground, but a point will be reached when the costs to extract these exceeds the price at which they could be sold as fuel. Eventually, a tiny amount of coal that is left—can you believe this?—might become so rare as to be a treasure, something people might frame and mount on a wall at home to show the fossil leaf of a tree that once lived and was turned into coal.

Economically recoverable means that a mining company can sell the coal it obtains at more than it cost to mine it. Physicists and mining engineers will also tell you that a fossil fuel has to be *energetically recoverable*, which means that the amount of energy in the fuel at the point of use is greater than the energy expended to obtain it and transport it to that location. This includes the energy costs of all indirect activities, such as pollution control resulting from mining, other production costs, and transportation.

There is a subtext in these chapters as well. If we have to move away from nonrenewable energy sources simply because they aren't renewable, what is stopping us from doing so? Is it politics, or big money, or our personal preferences, or something about the way our society is set up that forces us to use energy in certain ways, or what? I explore this to the extent that I can, but you may have some ideas about this, too.

We begin, then, with where we are now and what the consequences of our present condition are for us and for the world.

1

Oil

FIGURE 1.1 A modern oil-drilling ocean platform. Platform Holly, a few miles off the coast from Santa Barbara California, was installed in 1966 and has pro-duced oil since. *(Source: Linda Krop, Environmental Defense Center, Santa Barbara, CA)*

Key facts

- Worldwide, people use about 30 billion barrels of oil a year, which works out to 210 gallons per person. The worldwide total is expected to increase to 50 billion barrels a year—350 gallons per man, woman, and child—in the next few decades.

- In 2005, the United States used 28% of all the oil consumed in the world.

- In recent years, the United States consumed about 7.5 billion barrels of petroleum a year, dropping to 7.1 billion barrels 2008 (23% of the world's total consumption). More than 60% is imported; 17% of that is from the Persian Gulf.

- Two-thirds of all transportation energy in the United States comes from petroleum—2.2 billion gallons a day: 55% of this for ground transport of people, almost 36% for ground transport of freight, and just under 10% for air transport of both people and freight.[1]

- According to conventional estimates, at the current rate of use Americans will run out of oil in less than 50 years.

It's a stretch, but imagine you're an Eskimo living 1,500 years ago

It's around A.D. 500, and you're part of a small group of Eskimos struggling northeast in Siberia near the Bering Strait and crossing by boat into what is now Alaska. There you find other Eskimo groups whose lives are a struggle—living at the margin, barely enough food, hard to do anything but try to keep warm and figure out where the next meal will come from. This was the life of most Canadian Eskimos at that time, a struggle for existence.

But according to anthropologist John R. Bockstoce, an expert on Eskimo culture and Eskimo and Yankee whaling, you and your Eskimo relatives coming from Siberia, called the Birnirk culture, brought with you inventions for hunting. One of these was a harpoon made of bone and antlers that, like a modern whaling harpoon, would slide closed into the flesh of the whale and then lock in an open position when the whale tried to swim away. Your group also had kayaks, umiaks, and drag-float equipment and began using these devices to hunt whales. This led to a fundamental change in your lives. Whale meat and oil gave you so much more energy than your neighbors that your group did much more than simply hunt and think about the next meal. With the basic necessities of life—food and shelter—assured, people could use their surplus energy and time in more enjoyable ways: telling stories, painting pictures, singing—in other words, being "civilized" in the modern sense. Or if they were concerned that their food supply might dwindle, they could use that excess energy and time to acquire more territory, more food, more power—in other words, to wage war.[2] The ability of early Eskimos to obtain meat and oil from whales is analogous to our ability to get petroleum cheaply and easily from the ground. As long as it was available that way, we could while away our leisure time with video games, golf, travel, and whatever else we wished. But by now almost everyone understands that petroleum is a finite resource that will be used up pretty soon if we continue to rely on it as one of our major sources of energy. Moreover, it's equally clear that the use of petroleum, rather than declining, is going to increase, especially since the huge populations in China and India are rapidly increasing their ownership and use of automobiles.

Where does petroleum come from?

The fossil fuels—petroleum, natural gas, and coal—are just that, fossils. Coal was formed from the remains of trees and other woody plants, covered by soil and then buried deeper and deeper and subjected to heat and pressure, which converted their remains to mostly carbon, but with a fair amount of other elements that were part of the plants and the surrounding soil (for more on this, see Chapter 3, "Coal"). Petroleum and natural gas are believed to be the fossil remains of marine organisms.

All fossil fuels that we take out of the ground today were produced eons ago from the growth of photosynthetic organisms—algae, certain bacteria, and green land plants, organisms that can convert the energy in sunlight into energy stored in organic compounds, and do so by removing carbon dioxide from the atmosphere and releasing pure oxygen. The energy that fossil fuels contain is thus a form of solar energy, in most cases provided over many millions of years and stored since then.

Over time, much of the carbon from the carbon dioxide that algae, green plants, and some bacteria removed from the atmosphere was then sequestered—stored in the soils, rocks, and marine deposits, and prevented by various physical and chemical processes from returning to the atmosphere. When fossil fuels are burned, the sequestered carbon is released into the atmosphere as carbon dioxide (CO_2), which acts as one of the primary greenhouse gases.

Since petroleum and natural gas are not solids and thus are lighter than the rocks that surround them deep in the Earth (Figure 1.2), they tend to rise under pressure from the rocks and get trapped in geological pockets, like the one shown in Figure 1.3—although in some rarer situations the oil makes it to the surface, as it does in Southern California. Thus, the search for oil and gas is not random; petroleum geologists know which kinds of rock formations they are likely to occur in.

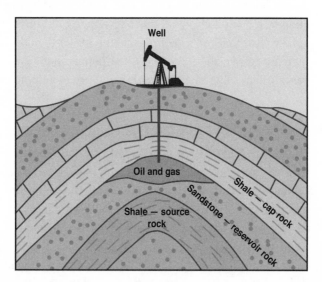

FIGURE 1.2 A typical location of oil and gas. Oil or gas rarely gets pushed right up to the surface, as it does at the La Brea pits in Los Angeles, famous for having trapped many ancient and extinct mammals whose fossils have become familiar. *(Source: D. B. Botkin, and E. A. Keller,* Environmental Science: Earth as a Living Planet. *New York, John Wiley, 2009)*

5-22-89

FIGURE 1.3   A natural oil seep along the California shore at Santa Barbara. Pressure from surrounding rocks has pushed petroleum up to the surface, where it flows into the Pacific Ocean, revealing itself by its bright reflection of sunlight. *(Courtesy of the University of California, Santa Barbara, UCSB Map & Image Laboratory, from the research collection of Prof. Jack E. Estes)*[3]

How much energy does petroleum provide?

In recent years, the United States consumed about 7.5 billion barrels of petroleum a year, dropping to 7.1 billion barrels 2008. More than 60% of petroleum is imported; 17% of this from the Persian Gulf.[4] Petroleum provides about 37% of the world's energy and 41% of the energy used in the United States,[5] most of which is used for transportation. The United States alone uses 8.4 billion barrels of oil a year. According to the U.S. Department of Energy, essentially all the energy used in transportation in the United States comes from fossil fuels,[6] and two-thirds of all transportation energy in the United States comes from petroleum: 2.2 billion gallons a day—55% (1.2 billion gallons a day) for ground transport of people, almost 36% (789 million gallons a day) for ground transport of freight, and just under 10% (210 million gallons a day) for air transport of both people and freight.[7] In contrast, petroleum provides only 1% of the electricity produced in the United States.[8] Most electricity in the United States is produced from coal, hydropower, and nuclear power. To keep things simple, think about petroleum as *the transportation fossil fuel.*

How much petroleum is there, and how long will it last?

These are straightforward questions, so one might expect them to have straightforward answers. Don't petroleum geologists and oil corporations know how much oil is in the ground, how much they can sell in a year, and therefore how long oil will last? Wouldn't this be a basic part of an oil company's business plan?

Unfortunately, it's not that simple. As economists and petroleum geologists will tell you, there is always more in the ground than you can get out, and the percentage you get out depends on how hard you want to work, or how much you are willing to pay, to get it. When oil was very, very cheap, around the first decades of the 20th century, it wasn't worth much to develop new technologies to get every last drop when the initial gusher and subsequent flow eased and the oil no longer flowed freely out of the ground. Today, we have many ways to push more of the under- ground oil to the surface or separate it from the rocks that hold it. So the answer to how much oil is in the ground is: It depends on what you are willing to pay.

As to the second question—how long will Earth's petroleum last?— economists will tell you that rather than being drained to the last drop, petroleum will eventually become so rare and so expensive to get out of the ground that it will no longer be useful as fuel. People may collect it, the way they collect other precious minerals, and display little jars of the black goo on their coffee tables as decorations and as evidence of their wealth. The real question therefore is not when every drop of oil will be gone but when it will no longer be economically worthwhile to extract it.

What will raise the price of oil and thereby make it worthwhile to try harder and harder to get it? One standard answer is that the price of oil will rise rapidly when peak production is reached—that is, when discov- ery of new oil declines. Another economic turning point is when the rate of supply drops significantly below the demand.

As petroleum reserves shrink, they get harder and harder to find

We're using more and more petroleum and finding less and less of it. Indeed, petroleum geologists suggest that we're going to run out of petroleum in the next few decades. History seems to be on the side of this viewpoint. In 1940, five times as much oil was discovered as consumed. Forty years later, in 1980, the amount of petroleum discovered just about equaled the amount consumed. And by the turn of the 21st century, world consumption of petroleum was three times the amount that was discovered.[9] Based on this history and our knowledge of the kinds of rocks where petroleum can be found, it seems likely that oil production in the United States will end in 50 years or at least by the end of this century, and world oil production soon after.

To understand how petroleum geologists think about these things and make calculations, you first need to understand the terms *resources* and *reserves*. A petroleum resource is the oil that can be extracted economically. A reserve is part of the resource, the part that, at the time it is evaluated, is judged to be eventually extractable both legally and economically. *Proven reserves* are those that have been determined to be legally and economically extractable right now. (The proven reserves idea leaves open the possibility that as prices for petroleum rise, it may become economically worthwhile to extract oil from reserves that are now considered too costly to use.)

Today, petroleum geologists estimate that the world's proven reserves are 1 trillion barrels (42 trillion gallons), and that total reserves—oil that eventually will be legally and economically accessible—are probably 2–3 trillion barrels. These estimates are based on a lot of geological knowledge as well as the location and size of existing oil wells. In fact, there is a wide range in the estimates of how many barrels of oil are now or soon will be considered proven reserves. For example, the U.S. Energy Information Administration reports values from 1–4 trillion barrels.[10]

In predicting when the oil supply will become a serious problem, petroleum geologists focus on the peak oil point—the time when one-half of Earth's oil has been exploited. This is usually projected to occur sometime between 2020 and 2050, although a variety of experts believe it has occurred already in the United States. The time of peak oil production is important because we can assume that when that point is reached, the price of oil will rise rapidly. The Energy Information Administration presents a range of estimates for the time of world peak oil production, from as early as 2020 to as far into the future as 2121.[11]

The implications are huge about how much time this gives the nations of the world to prepare for a planet without petroleum. Given the way most people and societies go about planning for events that they hope won't occur until far in the future, it seems likely that if peak oil production is expected to occur a century or more from now, little will be done to move away from fossil fuels in the next year or even the next decade, and when the time comes we'll all just muddle through. This will be unfortunate, because moving away from petroleum (and the other fossil fuels) is a good idea for reasons other than direct energy supply. For example, we could stop worrying about international conflicts over oil, avoid direct pollution from toxins given off by petroleum, and reduce the release of greenhouse gases. Those who place a high priority on a healthful, pleasant, and sustainable environment would therefore prefer to be told that peak oil is almost upon us, so that nations will be spurred to action.

For a more straightforward estimate of when the world will run out of petroleum, here are some numbers. Worldwide, people use about 30 billion barrels of oil a year (210 gallons a year per person). Conservatively—not taking into account the maximum potential increase in automobiles in China and India—worldwide consumption is expected to rise to about 50 billion barrels a year by 2020, which means that the whole world will use up today's proven petroleum reserves in about 20–40 years and use up the total estimated reserves in about 60 years. Since not all our many uses of petroleum may be readily adaptable to other fuels, this puts a lot of time pressure on all nations to get something going quickly to replace petroleum, especially for transportation.

However, there is another point of view, which is that conventional petroleum geologists greatly underestimate both the available amount of petroleum and how efficiently oil can be gotten out of wells. This viewpoint was well expressed in the *Wall Street Journal* op-ed piece titled "The World Has Plenty of Oil," by Nansen G. Saleri, president and CEO of Quantum Reservoir Impact, in Houston, and former head of reservoir management for Saudi Aramco.[12]

Mr. Saleri says that present oil mining technology gets only one-third of the oil out of a well; the rest clings to the rocks and is just held too tightly for current pumping methods to get it out. "Modern science and unfolding technologies will, in all likelihood, double recovery efficiencies," he writes. "Even a 10% gain in extraction efficiency on a global scale will unlock 1.2–1.6 trillion barrels of extra resources—an additional 50-year supply at current consumption rates."[13]

Mr. Saleri argues that rising prices for petroleum will fuel technological development that will increase extraction efficiency. Two major oil fields in Saudi Arabia are already yielding two-thirds, rather than one-third, of the oil out of the wells. Mr. Saleri writes that the total resources are 12–16 trillion barrels, not the 1 to 3 trillion barrels of conventional estimates, and that 6–8 trillion of these total resources are in conventional wells, the rest in "unconventional" sources, shale oil and tar sands, from which it is difficult and environmentally costly to get oil. Present attempts to recover oil from these unconventional sources are disrupting and polluting land (more about that later). Even with his optimistic assumptions, he estimates that peak oil production will be reached between 2045 and 2067—in 38–59 years.

Geography is against us

Unfortunately for most of us, petroleum reserves are not distributed evenly around the world. Quite the opposite; they are highly concentrated (Figure 1.4) and, worse yet, concentrated in parts of the world that, on the whole, are not the ones that use the most petroleum today but will likely require more in the future (Figure 1.5). The Middle East has 62% of the world's oil reserves (Figure 1.6); the rest of Africa 9.7%; South and Central America 8.6% (most of it in Venezuela and Brazil); the Russian Federation 6.6%. North America has just 5%, half of it in the United States.[14] So, as Figure 1.4 makes clear, oil reserves are extraordinarily concentrated geographically.

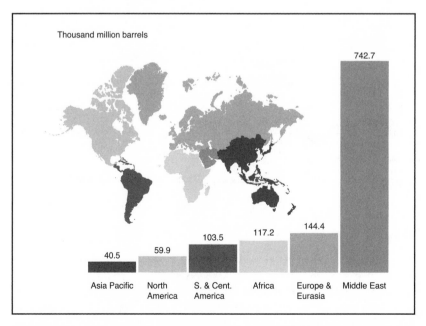

FIGURE 1.4 The world's known oil reserves (2006).[15] *(Source: BP Statistical Review of World Energy, June 2007; London, British Petroleum Company)*

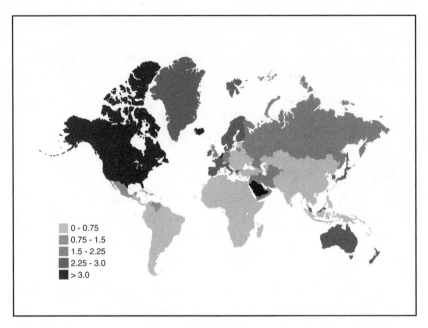

FIGURE 1.5   Oil consumption per capita (metric tonnes, 2006). Compare the consumption with known reserves. *(Source: BP Statistical Review of World Energy June 2007; London, British Petroleum Company)*

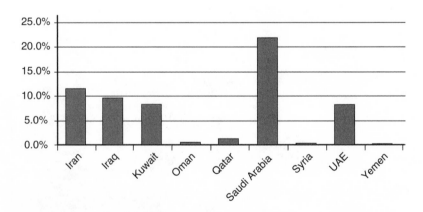

FIGURE 1.6   Middle Eastern nations have 62% of the world's available oil. Most of this is in five nations: Saudi Arabia (with more than 20%), Iran, Iraq, Kuwait, and the United Arab Emirates.[16]

It would be naive to think that the lopsided geographic distribution of petroleum will not continue to create international conflicts. As long as the United States and other countries without vast oil reserves continue to depend so heavily on petroleum, these conflicts are likely to increase, which is all the more reason to turn to other sources of energy as soon as possible.

Although the Middle East dominates world oil reserves, most of that oil goes to Europe, Japan, and Southeast Asia, whereas the United States imports a lot of oil from Canada, Mexico, and Venezuela (Figure 1.7). Obviously, the more oil the United States imports, the more vulnerable its economy is to the reserves in other nations and to political and environmental events that limit or prevent this importation. Given the importance of abundant energy for a vibrant economy and society, greater energy independence is an important goal, but for petroleum this is not and will not be possible for the United States.

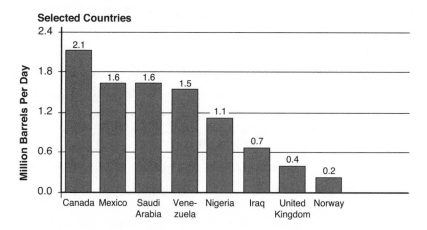

FIGURE 1.7   Where the United States gets its oil. *(Source: Energy Information Administration, 2008)*[17]

Where might new oil reserves be found?

Recent discoveries of oil have been primarily in the Middle East, Venezuela, and Kazakhstan.[18] Ironically, global warming may change this, since less ice in the Arctic may mean more opportunities for oil exploration where it was difficult before. Also, while at present drilling for oil in Arctic waters is mostly limited to a depth of 300 feet and in some cases 2,000 feet, new ships will make it possible to drill for oil in water 12,000 feet deep.[19] One estimate suggests that 400 billion barrels of oil may be found in the Arctic oceans.[20]

Although this is a lot of petroleum, at current rates of use it would add only eight years to the time we have before the world runs out of oil.[21] And there's a potential downside: More global warming provides more sources of oil—for example in the Arctic, which produces more greenhouse gases, which lead to more global warming.[22] Then, too, there is already plenty of concern about oil spills and their effects on ocean ecosystems, sea and shore birds, and fisheries, and the ability to drill much deeper in a much larger area increases the risk of drilling-caused spills.

Two unconventional sources of oil: oil shales and tar sands

As explained earlier, petroleum under pressure from underground rocks fills pockets in the rocks. But in addition, some muds trap petroleum as they form into shales, resulting in a dense rock filled with oil. The oil is tightly bound within the rock and can be released only if the shale is heated to 900°F. At this temperature, a ton of shale may yield as much as 14 gallons, and three tons of shale would be needed for each barrel of oil. Heating three tons of rock to 900° takes a lot of energy and leaves behind a lot of crushed rock. Much of this rock is obtained from surface mines, and even more energy is needed afterward to restore the damaged land—restore it as much as possible, that is. Not only are oil shales a highly polluting energy source, destructive to the land, but also their net energy yield is low compared to conventional sources of oil.

Tar sands (sometimes also called oil sands) are geologically similar to oil shales, but the petroleum impregnates sand or clay rather than mud. Again, the petroleum is so completely mixed with the inorganic material that one can't pump the oil out. The sand has to be mined, primarily by strip mining, and then washed with hot water. As with oil shales, a mess remains—in this case dirty water as well as tons of sandy rock. Tar sands are said to yield as much as one barrel for about every two tons processed.

Those who believe there is a lot more oil out there than 1–3 trillion barrels are basing their estimates partly on what could be gotten from oil shale and tar sands. An estimated 3 trillion barrels of oil exist in oil shales and about the same in tar sands. Together, these massive but difficult-to-use sources could triple the amount of oil available, if all of it could be recovered.

Much of the world's known tar sands and oil shales are in North America. The United States has two-thirds of the known world oil shale, and it is estimated to contain 2 trillion barrels of oil. Some 90% of U.S. oil shale is in the Green River formation underlying parts of Colorado, Utah, and Wyoming and extends over 17,000 square miles, an area larger than Maryland.[23] Canada has an estimated 3 trillion barrels of oil in tar sands, most of it in a single huge area near Alberta now called the Athabasca Oil Sands. Since so much energy is required to get the oil out of these rocks, the net yield would not be nearly as great as from conventional oil wells. Still, the government of Alberta states that tar sands yield six times the amount of energy required to process them.[24]

Oil shales and tar sands are already causing major environmental controversies, since so much oil exists in them, and since mining and refining it are so polluting. Mining the 2 trillion barrels of petroleum from U.S. oil shales would leave behind 9 trillion tons of waste rock—an amount equal to the weight of 24 *million* Empire State Buildings. To put this into perspective, in 2007, all the freight transported in the United States weighed 21 billion tons. So it would take all the freight transportation available in the United States about 424 years to move that much waste rock.[25]

Three tar sands mines are operating today: Suncor (opened in 1967), Syncrude (since 1978), and Muskeg River of Shell Canada (opened in 2003). They are producing 1 million barrels a day[26] and have affected 120 square miles.[27] Mining Athabasca Oil Sands takes 2.2 to 5 barrels of water for every barrel of oil.[28] Water used for this processing comes from the Athabasca River, which starts in the beautiful Canadian Rockies as the outflow from the Athabasca Glacier. The government of Alberta states that only 3% of the average annual outflow of the glacier is required to process the sands,[29] but environmental groups estimate that it will require a quarter of Alberta's freshwater.[30] This water would end up in holding ponds, contaminated by toxic chemicals from the mining and processing: mercury, arsenic, and a variety of organic compounds that are carcinogenic.[31]

Effluents from present tar sand operations are being blamed for human and wildlife ailments,[32] and the holding ponds present an even greater hazard. According to Professor David Schindler of the University of Alberta, a leading aquatic ecologist, "If any of those tailings ponds were ever to breach and discharge into the river, the world would forever forget about the *Exxon Valdez*."[33]

Currently, oil production from Athabasca Oil Sands costs between $15 and $26 a barrel, compared with about $1 per barrel from Saudi Arabia's wells. But when oil prices exceeded $130 a barrel in 2008, mining those tar sands began to sound like economic sense—except for the pollution (at the time of this writing, oil is $72 a barrel).

Oil shales are not yet in commercial development, but Shell Oil Corporation has invested many millions of dollars in attempts to develop this petroleum source.[34] The near future will bring a major battle over North American tar sands and oil shales since they offer huge profits at great environmental costs.

Growing worldwide competition for a dwindling resource

International competition for petroleum is growing, in large part because rapidly rising standards of living in India and China are leading to a greater number of automobiles. India now has 5.4 million vehicles, up 500% in just 20 years.[35] China has 34 million registered motor vehicles.[36] In 2006, sales of personal autos rose 30% in China, to 5.8 million,[37] and China's total vehicle sales reached 7.22 million. To put this into perspective, this is close to half the number of cars sold in the United States in 2007 (about 16 million).[38] In 2003, China became the world's fourth-largest automobile-producing nation, behind only the U.S., Japan, and Germany.[39] This increased competition alone is enough to push petroleum prices up. And they're going to go even higher. The cost of generating electricity with oil (and with natural gas) in the United States has been rising sharply. Domestic electricity cost 20% more in 2006 (the most recent date for which data are available) than in 1995.[40]

If supplies are dwindling, why watch petroleum go up in smoke?

On May 15, 2007, the *Wall Street Journal* reported that Aramco, a highly profitable state-run Saudi oil giant, had signed a huge deal with Dow Chemical. Why would the world's largest producer of fuel oil be interested in making a deal with a chemical company? Since petroleum is an excellent base for many artificial chemicals, a large number of very popular and very profitable products—including most plastics—that most of us would be unwilling to do without are made with them. According to the *Wall Street Journal,* the Aramco-Dow agreement is supposed to lead in 2013 to a joint venture that will build plants to produce 7 million tons a year of these chemicals.[41] And by the end of May 2008, Dow Chemical announced that it would have to raise the price of its petrochemicals 20% because of the rising price of crude oil.[42] Why waste whatever petroleum we have left by burning it all up fast as fuel? Why not use alternative energy sources and save petroleum for other important purposes that use much less of it?

Environmental effects of petroleum

Petroleum causes pollution at every stage, from mining and recovery to refining, transporting, and using it as fuel. Drilling wells can cause direct pollution via oil spills. Drilling also often involves injecting watery liquids into the wells; later released as drilling muds, these cause their own toxic pollution.

The notorious *Exxon Valdez* oil spill taught us that transporting oil by tanker ships can lead to disaster. Transporting oil by pipeline or truck can also lead to spills, because pipes break and trucks sometimes have accidents.

Crude oil—oil as it comes out of the ground—is many chemicals mixed together, and these must be separated into gasoline, kerosene, diesel fuel, heating oil, and heavier materials. This is what a refinery does: Like a giant chemistry set, it heats crude oil and separates its chemicals according to their density. The strong odors that make passersby wrinkle their noses are petroleum chemicals that the refinery has released into the environment—chemical pollutants. Travelers nearing the end of the New Jersey Turnpike on their way to the tunnels into New York City know exactly what I'm talking about.

These are just an indication of the potential for refineries to leak chemicals into the air, soil, and groundwater; to suffer accidental fires and breakages that produce more pollution; and to create sites that are heavily toxic for future generations.

Effects of the Exxon Valdez Alaskan oil spill are still with us

On March 24, 1989, the tanker *Exxon Valdez* spilled 10.8 million gallons of crude oil into Prince Edward Sound, Alaska. Although it was not the largest spill ever, the oil slick extended over 3,000 square miles and inflicted heavy damage. The wildlife affected have been estimated to include 250,000 to 500,000 seabirds, at least 1,000 sea otters, about 12 river otters, 300 harbor seals, 250 bald eagles, and 22 killer whales.[43] Nearly 19 years later, the spill still affects Alaska's fisheries, and lawsuits over its effects include the Alaskan Eskimos' $2.5 billion suit for damages.[44] Costs of this kind are not usually counted in tallying the total costs of petroleum.

Petroleum exploration versus conservation of endangered species

The Arctic National Wildlife Refuge is a classic example of the conflict between the search for more petroleum and the conservation of wildlife and endangered species. The refuge is in beautiful country. It was established in 1960 and expanded in 1980 to cover 19 million acres, larger than the combined area of Massachusetts, New Hampshire, and Vermont (Figure 1.8, top). It is the primary breeding ground for 123,000 caribou of the Porcupine herd (named for the Porcupine River [Figure 1.8, bottom]) and is also a major wintering ground for this population.

The refuge also contains an estimated 10 billion barrels of oil. How much of this could be recovered is uncertain—conservative estimates are about 3 billion barrels. The possibility of drilling for oil in the refuge was remote until the 21st century; the George W. Bush administration pushed for it, arguing that it would help make the United States more energy-independent. But the United States has been using about 7.5 billion barrels of oil a year, so at best all the oil in the Arctic National

Wildlife Refuge would buy the U.S. less than a year's worth of oil. At the time of this writing, neither the Obama administration nor Congress has made any decisions about drilling there.

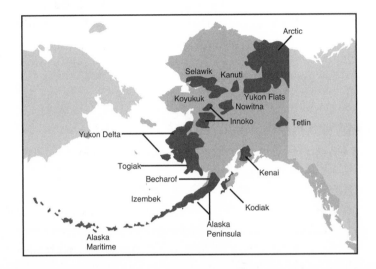

FIGURE 1.8 Map of Alaskan National Wildlife Refuges (top). Each dark area is a refuge. The Arctic National Wildlife Refuge, the one discussed in this book, is listed as "Arctic" at the top right. Caribou within the refuge (bottom). *(Courtesy of the U.S. Fish & Wildlife Service: http://arctic.fws.gov/caribou.htm. See also http://arctic.fws.gov/pdf/ispch.pdf)*

Here are some of the other ways that petroleum pollutes. Burning petroleum pollutes the air, creating health problems and damaging plants and wildlife. Among the primary petroleum-generated air pollutants are ozone, nitrogen oxides, and particulates. Also, pipelines and storage tanks leak. In 2001 a rifle bullet punctured the Trans-Alaska Pipeline, resulting in a small but nonetheless damaging spill. Among the good news is that although the 2002 Alaska earthquake ruptured the earth under the pipeline, the line stayed intact.

The bottom line

- Known petroleum sources will run out in less than 50 years (according to conventional analysis) or perhaps in 100 years or so (unconventional analysis).

- Whatever the exact time when petroleum runs out, we have a choice: We can devote a large portion of our time, resources, and energy to seeking new oil and improving extraction efficiency, or we can seek sustainable and cleaner energy sources.

- Petroleum is one of the three most polluting energy sources (the other two are nuclear power and coal). The potential for pollution will increase as conventional oil sources run out and the world turns to the unconventional sources: tar sands, oil shales, and deep ocean drilling.

- In an ideal world, the search for new energy sources would move away from petroleum, but so much money can be made from obtaining and selling crude oil that oil development will likely continue in the short term despite increasing pollution and increasing knowledge of its health and environmental effects.

2

Natural gas

FIGURE 2.1 A large tank, center, falls into a sinkhole near Daisetta, Texas. The sinkhole was caused by mining beneath the surface for natural gas. *(Source: © AP Photo/KHOU-TV, Bobby Bracken)*[1]

Natural gas is capturing the public's imagination, especially since many well-known people, from T. Boone Pickens to Nancy Pelosi, have praised it as the cheap, clean, and abundant fuel for the future. But new technologies for drilling for natural gas will have large environmental effects, like this sinkhole in Texas (Figure 2.1) where large amounts of water were pumped into the ground to release gas from the rocks below.

Key facts

- Almost a quarter of all energy used in the world and in the United States is provided by natural gas.

- At current rates of use, the world's known reserves of natural gas will last about 60–65 years, to about 2070—longer than petroleum.

- But if the United States seeks energy independence and obtains natural gas only from within U.S. borders, our reserves will be exhausted in just 16–17 years, even at the 2006 rate of use.

- In contrast to petroleum, natural gas has played a small role in powering vehicles. In the United States, about 20% of it is used to heat homes and about one-quarter of it to generate electricity (some of which, of course, may be used to power vehicles, especially trains). Most of the rest is used for office buildings and manufacturing.

Is Utah leading the way to the future fueling of cars and trucks?

It may seem surprising, but until recently natural gas was considered an unusable waste and was burned off at oil wells and refineries since it was difficult and dangerous to transport. That problem was solved by long-distance pipelines and by transporting liquefied natural gas by ship, train, and truck. Utah was the first of the 50 states to make a major push to use natural gas for automobiles, and by the end of August 2008, when gasoline was surging above $4 a gallon, the idea was a big hit. Unusual everywhere else at the time, compressed natural gas was said to be widely available for automobiles in Utah, and it was cheap, about 87 cents for enough to run a car the same distance as a gallon of gasoline.

The idea had a lot of support—"Utah shows that the technology is here and the fuel works and the fuel is better than foreign oil," said T. Boone Pickens, who has been promoting natural gas and pushing for a big transition from petroleum to alternative energy by using natural gas as the major intermediary. When Jon M. Huntsman Jr. was governor of Utah (he's now the American ambassador to China), he spent $12,000 of his own money to convert the state SUVs he drove so they could run on natural gas. "We can create a model that others can look to," he said.

"Every state in America can make this a reality."[2] T. Boone Pickens's energy plan, available on his website, states that "natural gas is our country's second largest energy resource and a vital component of our energy supply. Ninety-eight percent of the natural gas used in the United States is from North America. But 70% of our oil is purchased from foreign nations." It goes on to say that "domestic natural gas reserves are twice that of petroleum. And new discoveries of natural gas and ongoing development of renewable biogas are continually adding to existing reserves."[3] Congress got into the act as well, with Nancy Pelosi, Speaker of the House, saying on "Meet the Press" that "you can have a transition with natural gas that is cheap, abundant, and clean."

It sounded like a good idea. But at the same time in Utah there were hints that natural gas might not be the easy panacea that Pickens, Pelosi, and Governor Huntsman were suggesting. At rush hour, stations selling the gas were finding that the pressure was running low, and sometimes people couldn't find enough to fill more than half a tank. The supply couldn't meet the demand—and that was when only 6,000 of the 2.7 million vehicles registered in Utah were running on natural gas. Not only was there a shortage of filling stations selling natural gas in the state, but also a car's tank could hold only enough compressed gas to take you about half as far as a tank of gasoline.

Still, natural gas was promoted across the country, and plans were developing rapidly to increase the number of natural-gas-powered cars through government subsidies. The *New York Times* reported in August 2008 that "a proposal on the ballot in California this fall would allow the state to sell $5 billion in bonds to finance rebates of $2,000 and more to buyers of natural gas vehicles. Legislation has been introduced in Congress to offer more tax credits to producers and consumers and mandate the installation of gas pumps in certain service stations, with the goal of making natural gas cars 10 percent of the nation's vehicle fleet over the next decade."[4]

How much energy does and could natural gas provide?

At present, natural gas provides 22% of the energy used in the United States, which makes us one of the world's heaviest users of this form of energy. About 20% of the natural gas we use is for heating homes, and about one-quarter generates electricity (Figure 2.2). Transportation uses 28% of America's energy, and of that 28%, hardly any—one-tenth of a percent—comes from natural gas. (Although, to make things more complicated, some of the electricity generated from natural gas powers vehicles, especially trains.)

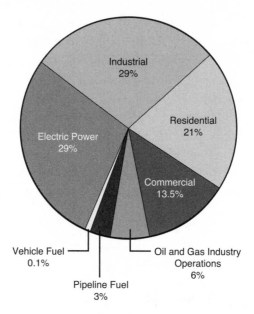

Figure 2.2 U.S. uses of natural gas in 2008. Vehicle use was only 0.1%. *(Source: Energy Information Administration, Natural Gas Annual 2008 [August 2009])*

Could natural gas really power as many as 10% of all U.S. vehicles within ten years? Or could the beginning of problems with the natural gas supply for automobiles in Utah be a hint as to what the future might bring for all of us? I began to look into the numbers about natural gas, as I had done for all other forms of energy, and here's what I found.

To learn about natural gas's potential, I contacted Ray Boswell, manager of Methane Hydrate R&D Programs, U.S. Department of Energy, National Energy Technology Laboratory, in Morgantown, West Virginia.[5] He provided the basic information shown in Figure 2.3, which in turn enabled me to make the calculations in Figure 2.4. The shocking result is that *with U.S. natural gas independence—that is, using only natural gas obtained within U.S. borders—even at the 2006 rate of use, the readily available gas within the United States would be exhausted in one year; that plus what is called "technically recoverable" would be gone in less than a decade; and what is termed "unknown but probable" would last about a century.*

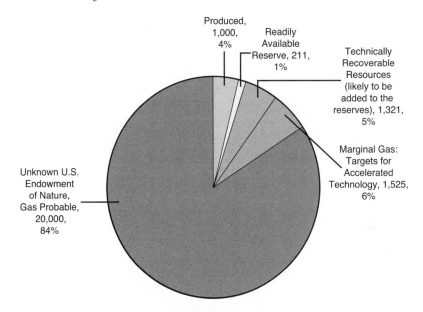

FIGURE 2.3 America's natural gas resources, in trillions of cubic feet. *(Source: Ray Boswell, U. S. Department of Energy)*

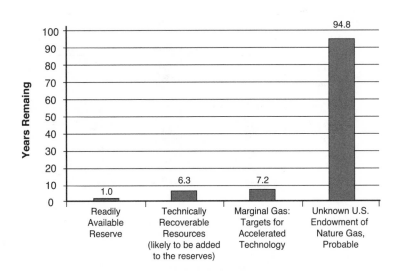

FIGURE 2.4 How long would U.S. natural gas last? This graph shows projected years remaining for "readily available," "technically recoverable," and "unknown but probable" natural gas reserves and resources at 2006 rates of use. *(Source: Ray Boswell, U. S. Department of Energy)*

Suppose we went all the way with the suggestion by T. Boone Pickens, Nancy Pelosi, and Governor Huntsman, and suddenly, tomorrow, all
automobiles and light trucks were able to run on natural gas. As we discuss later (see the chapter on transporting ourselves and our stuff), of all
transportation methods, cars and light trucks are the biggest petroleum
guzzlers, using 63% of transportation energy in the U.S. This change
would increase the total percentage of energy provided by natural gas to
just under 40%, our nation's "readily recoverable" plus "technically feasible" supplies would last less than 6 years, and the "unknown but probable" resources alone would get us another 8 years. Note that this
continues to assume that all other use will remain at 2006 levels. (Use
will most likely rise with the growth of America's population, but could in
fact decline with increased efficiency of cars and light trucks and an
increasing number of electric vehicles whose energy comes from sources
other than natural gas.)

In July 2009, the Department of Energy announced that the estimated U.S. gas reserves in the category "marginal gas: targets for accelerated technology" were 35% larger than previously estimated.[6]

Also in 2009, the Potential Gas Committee issued a report that cate-
gorized natural gas somewhat differently and did not include the
Department of Energy's "unknown but probable" (Figure 2.5).

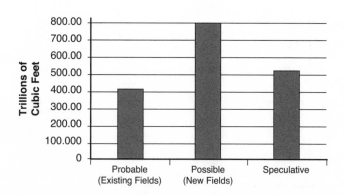

FIGURE 2.5 Another estimate of U.S. natural gas reserves, 2008.

We can also calculate how many years U.S. natural gas reserves
would last based on the Potential Gas Committee's estimates and the use
at 2006 rates with the addition of providing fuel for all cars and light

trucks (Figure 2.6). The result is a total of 11 years before the U.S. runs
out of natural gas.

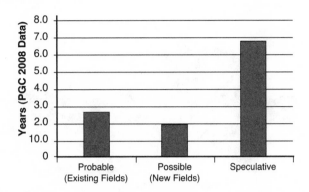

FIGURE 2.6 Years remaining of U.S. natural gas, using PGC 2008 estimates
and assuming that use is at 2006 rates and that all cars and light trucks use
only natural gas.

The take-home story is that an energy-independent America could
fuel its cars and trucks only with natural gas for less than 20 years, unless
we consider the most speculative and environmentally destructive natural gas sources. I am not a petroleum geologist, or a geologist of any kind,
so I cannot tell you how good the chances are that the "unknown but
probable" is actually out there. My correspondence with Ray Boswell
and others suggests, however, that the chances are good, and that if we
are willing to take a risk, we might enjoy natural-gas energy independence for as long as a century. This would require sizable upscaling of
exploration for this fuel, and the question is not simply whether we want
to take this risk for 80–100 years of energy independence but whether
we *have* to take this risk. What about the environmental effects of natural gas? It is the cleanest fossil fuel, but is it the *completely* clean fuel that
is being suggested?

World use of natural gas

Worldwide use and availability of natural gas are much the same as in the
United States. Natural gas provides 26% of the energy used worldwide,
slightly more than the 24% provided by coal but only about two-thirds
the amount provided by oil. And at present rates of use around the

world, recoverable natural gas will last about 60–65 years, according to
the U.S. Energy Information Agency (see Figures 2.7 and 2.8).

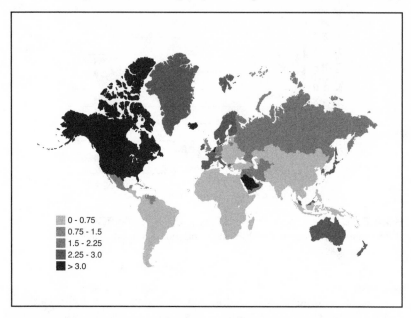

FIGURE 2.7 Worldwide per-capita consumption of natural gas in tons per
person. The U.S. and Canada are among the greatest per-capita users of natural gas. *(U.S. EIA. International Energy Annual 2006)*

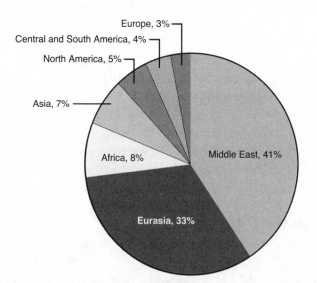

FIGURE 2.8 World reserves of natural gas as of January 1, 2009. *(Source:
U.S. Energy Information Agency. Redrawn by DBB 31 Jan 10.)*[7]

Global use of natural gas equals about 30 trillion kilowatt-hours a year. By comparison, Hoover Dam generates 4 billion kilowatt-hours a year, so the total world use of natural gas equals the energy produced by 7,500 Hoover Dams. The United States uses 6,463 billion kilowatt-hours per year from natural gas, which equals the energy from 1,615 Hoover Dams. (This is also equal to the energy in 5 billion tons of coal, which we discuss in the next chapter.)

Hope in the deep seas

In addition to the conventional sources of natural gas, several unconventional sources may provide large amounts of gas fuels. These sources are gas hydrates, coal-bed methane, and shale-bed methane. As I understand it, these possible sources are part of the "unknown but probable" resources but may even exceed those estimates.

Gas hydrates are frozen forms of organic gases, mainly methane, and are often referred to simply as methane hydrates. In a kind of water-ice matrix, they are buried in the ocean below 3,000 feet, where the temperature and pressure are sufficient to freeze the gas, and also in organic material in permafrost, where their emissions in small amounts are called *marsh gas*. Ocean deposits of methane hydrates were discovered only about 30 years ago, and only rough estimates of their quantity are available, but these suggest that methane hydrates might double, or even more than double, the total amount of energy available in all other known fossil fuel deposits—coal, oil, and natural gas (Figure 2.9). The problem is that methane hydrates are difficult to release from their ice matrix and make usable. A start has been made, but mining this gas is still in the experimental stage, with several test wells in Canada in development by the government of Japan.[8]

Coal-bed methane

We discuss coal in detail in Chapter 3. In brief, coal is formed when woody plants—trees and shrubs—die, are buried in wet ground, which limits the kinds of decay, and are later buried deep by new deposits above them. Heat and pressure from the newer deposits above convert the dead wood first to peat, then to lignite, and then to coal. Each step increases the amount of fuel and decreases impurities, including water. During this process, a lot of methane is produced, largely by the activity of certain kinds of bacteria that live only in the oxygenless environments of material soaked in water. This methane is released naturally in relatively small amounts by the heat and pressure, bubbling to the surface as swamp gas. But this natural gas can also be mined, and estimates are that there may be a lot of it available. According to one estimate, more than 20 trillion cubic meters of coal-bed methane may exist in the United States, of which about 3 trillion cubic meters could be mined economically today—about a 5-year supply at current rates of use of natural gas.[9] One drawback is that mining this gas will no doubt cause environmental damage similar to some forms of coal mining (discussed in the next chapter).

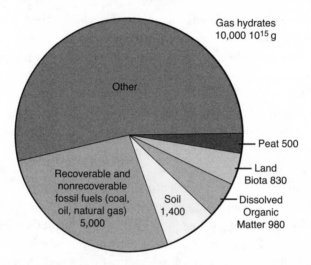

FIGURE 2.9   Methane hydrates may be the largest source of organic carbon in the Earth. *(U.S. Geological Survey, Gas [Methane] Hydrates—A New Frontier, Dr. William Dillon, Keith Kvenvolden, USGS)*

Natural gas from shale

Shale is one of the most common kinds of rock in the United States, readily found in 23 states (Figure 2.10). It forms the reddish earth common on the coastal plains of New Jersey and other states that front the Appalachian Mountains, and also out west in such states as Oklahoma and Texas. Like methane hydrate, gas from shale has captured a lot of attention and interest. One of the first areas of focus is Barnett Shale near Fort Worth, Texas. In mid-2008, 7% of U.S. natural gas production was said to be coming from this one formation.

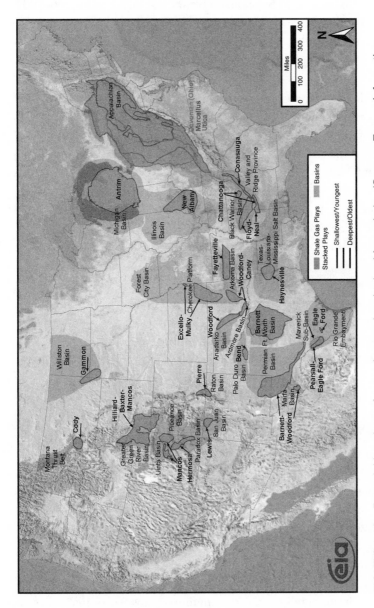

FIGURE 2.10 Potential locations of natural gas to be obtained from shale. *(Source: Energy Information Administration, May 28, 2009)*

In April 2008, the *Wall Street Journal* reported that estimates of the amount of gas that might be obtained from shale varied widely: "In 2002, the U.S. Geological Survey estimated there may be 1.9 trillion cubic feet," the article said, but "earlier this year, Terry Engelder, a Pennsylvania State University geosciences professor, made what he called a conservative estimate of 168 trillion cubic feet. His estimate has yet to be confirmed. By comparison, the U.S. consumed 23.05 trillion cubic feet last year, according to the Energy Information Administration."[10]

Four months later, in August, 2008, estimates had increased to "842 trillion cubic feet of retrievable gas in shales around the country, enough to supply about 40 years' worth of natural gas, at today's consumption rate," according to an article in the *New York Times*.[11]

Although it has been known that shale contains natural gas, until recently the technology was not available to retrieve it. The new technology to get at this gas involves drilling very long wells that lie horizontally within the rock, rather than descending vertically through it. Hot water is then pumped into the horizontal wells, and this fractures the rock, releasing the gas.[12] The reality is that this is a new form of strip mining, with all the environmental problems that result from that method, which I will discuss in more detail in the chapter on coal.

In all the excitement about natural gas as *the* solution to our energy problem, such consequences seem to be overlooked. However, Kate Sinding, a lawyer for the Natural Resources Defense Council, was quoted in the *New York Times* warning that intensive use of water in recovering gas from shale could pose an environmental threat, especially to local and regional water supplies. And an article in the *Wall Street Journal* showed a sinkhole the size of a football stadium (see this chapter's opening photograph) that was created in Texas by drilling for oil and injecting water into the ground to obtain natural gas from shale. That article stated that in 2006 alone more than 280 billion gallons of liquids, mostly water, had been injected into the ground as part of mining.[13] That's as much water as 7.8 million Americans use on average in a year.

According to another article in the *Wall Street Journal*, "federal regulators, environmentalists, and community groups worry that lax oversight is allowing some of the water—which can be ten times as salty as seawater and often contains oil, heavy metals, and even radioactive material—to escape from underground reservoirs. That could lead to the contamination of underground drinking-water supplies, the pollution of soil and surface water, and more sinkholes as underground structures are eroded."[14]

In short: If we are going to go the route of gas from shale, we had better expect a lot of pollution, similar to the effects of strip-mining coal.

As you read throughout this book, I believe that the solution to our energy problem will involve a variety of sources. Natural gas is the best and cleanest of the fossil fuels and, therefore, the one fossil fuel that we should emphasize. But to what extent and for what uses? Before making any decisions about that, we have to explore the other sources of energy, and that is what we do in the next chapters.

The bottom line

- Of the fossil fuels, natural gas has become the darling, with famous businessmen and politicians promoting it as the clean, cheap, and abundant fuel of the future.

- Natural gas is the cleanest to burn of the fossil fuels and is especially valuable for such uses as urban transportation, where it is important to minimize local chemical and particulate pollution of the air, and for running small turbines for peak power production.

- U.S. natural gas reserves cannot provide energy-independence for very long, because natural gas from traditional wells is limited within the United States. The largest potential sources of natural gas are in the deep sea, in coal beds, and in shale, and obtaining natural gas from these new sources is challenging and can create large-scale environmental problems.

- Natural gas will continue to be important for cooking and space heating, but despite what you may hear, obtaining enough of it for transportation and other uses for America's growing population and for the world in general will not be easy or cheap.

- Even if the kinds and rates of use remain at 2006 levels, readily available natural gas from wells that pollute little would last less than a decade. And if it becomes a major fuel for cars and light trucks, U.S. natural gas from minimally polluting wells will run out even sooner.

3

Coal

FIGURE 3.1 A coal-gas streetlamp in 19th-century Washington, DC.[1] The crossbar just below the glass lamp is a place for the lamp lighter to rest his ladder. *(Source: Library of Congress, the Brady Collection, LC-DIG-cwpb-03640)*

Key facts

- Coal provides nearly 60% of electricity and 25% of total energy in the United States today.

- There are 476 coal-fired power plants in the United States. Advocates propose about 150 more, some of which are in construction; others have been approved by the government.

- Coal use is increasing rapidly around the world and reaches a new record high each year in the United States.

- China is rapidly building more coal-fired power plants and may soon catch up with the United States in both the amount of electric power produced from coal and the amount of CO_2 released in the process.

- At current and projected rates of use, coal reserves will last another 150 to 300 years.

- Coal, along with nuclear power, is the dirtiest form of energy, and coal mining and burning have a long history of causing damage to the environment and to human health. Coal mining has destroyed towns, landscapes, mountaintops, rivers, and streams. Coal burning releases toxic elements such as lead, mercury, and arsenic and is a leading cause of air pollution and acid rain.

This coal comes with laundering instructions

Just before Christmas 2007, a consortium of some of the world's major electric utilities and coal mining companies—American Electric Power, Peabody Energy, Rio Tinto Energy America, and Southern Company, along with Australian, British, and Chinese companies—announced that a new kind of coal-fired electric power plant would be built in Mattoon, Illinois, which would provide the cleanest power in the world. The planned new power plant was part of the U.S. federal FutureGen program, announced in 2003 by President George W. Bush. The claim was that its power generation would be clean because instead of burning coal directly, the energy in the coal would be converted to hydrogen gas, and the carbon dioxide released from the burning coal would be buried deep in the ground rather than released into the atmosphere. The Illinois plant was planned to generate 275 million watts and cost $1.4 billion;

$1 billion of which would come from the federal government. It was supposed to begin operating in 2015.[2]

Transferring the energy in solid coal to a gas is not new—it was invented in the 18th century. In the early 1800s, before the invention of the electric light, coal gas fueled streetlamps in London (the first to have this kind of lighting) and Philadelphia (America's first coal gas streetlights), Boston, Washington, D.C., and New York (Figure 3.1). At that time, street lighting from coal gas was new and considered a great advance. Coal gas began to light homes by 1830.[3]

Although coal gas technology has been around for almost two centuries, the FutureGen power plant in Illinois was to be new in two ways: It would be the first operational plant to bury carbon dioxide produced from burning coal, and it would combine this with the latest method to make coal gas, called an *integrated gasification combined cycle (IGCC)*. The IGCC method of coal gas production is used in only four power plants around the world: Puertollano, Spain; Buggenum, the Netherlands; Terre Haute, Indiana; and Polk County, Florida.[4]

"There is no project in the world that can move near-zero-emission power and CCS (carbon capture and storage) further or faster than FutureGen at Mattoon," Senator Dale Righter, Republican of Illinois, said. "Today I could not be more proud.... I look forward to the next step where we make this promise of economic evolution a reality." However, FutureGen's electricity is estimated to cost three times as much as energy from a conventional coal-fired power plant. As we will see in later chapters, this raises the cost beyond electricity from wind power and approaches that from solar power.[5]

And will it work? And even if it does work, will it be worth $1 billion of the taxpayers' money? And is it the best way to get cleaner, more secure, sustainable energy? This chapter can help you to make a decision about FutureGen and the future of coal in general.

What exactly is coal?

Coal is fossilized land plants that have been buried deep in the earth for millions of years. Although the intense heat and pressure at those depths have converted the vegetation into a hard, black material that is primarily pure carbon, you can still see the leaves and stems of plants in the fossils.

Most coal was formed during the Carboniferous and Permian periods, 363–245 million years ago, and in the Cretaceous period, 146–45 million years ago, when wetlands were widespread.[6] Dead vegetation that lay buried in large wetlands could not decay, because the water prevented oxygen in the air from reaching it.

There are four kinds of coal, depending on how long the plant material has been underground, at how high a temperature, and under how much pressure. Anthracite (hard coal) is the hardest and best coal fuel because it is 86% to 98% carbon and has the fewest impurities. Bituminous (soft coal) is the most abundant but is softer than anthracite, has a lot of impurities, and is only 69% to 86% carbon. Subbituminous (medium-soft coal) is even softer and has more water and impurities. Lignite (brown coal) is the worst stuff, very soft, 70% water, and less than 30% carbon. Of the four, anthracite, the cleanest and with the least water, is the best for home heating. The others are used primarily to produce electricity, and bituminous coal is also used to make coke, which in turn is used with iron to make steel.[7]

It is estimated that the Earth contains approximately 1 trillion tons of coal. The energy stored in this is somewhere between 4.76 and 7.64 million billion kilowatt-hours, a number hard to imagine. Here's a try: Hoover Dam, as mentioned earlier, generates 4 billion kilowatt-hours a year, so the world's total coal reserve contains the amount of energy that would be generated by 1 to 2 million Hoover Dams running at full capacity for one year, or, if you prefer, 1 Hoover Dam running for 1 to 2 million years, or about 100 Hoover Dams running for 10,000 years—about the length of time that human civilization has existed on Earth. No matter how you look at it, there is a lot of energy stored in coal, thanks to woody plants that over hundreds of millions of years did not decay but just lay quietly, buried and stored.

Like oil and natural gas, coal is distributed unevenly around the world, but coal is found in more locations than the other fossil fuels—70 nations have recoverable coal, but 10 nations currently produce 95% of the world's coal (Figure 3.2).

The United States has a lot of coal—it is the second-largest producer, behind only China in its mining. Coal in the United States is concentrated in the central and southern Appalachian Mountains of the East and in the western Great Plains, with some in the Midwest (Figure 3.3).

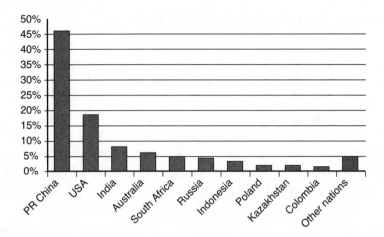

FIGURE 3.2 Although there is a lot of coal around the world, these ten nations mine most of it.

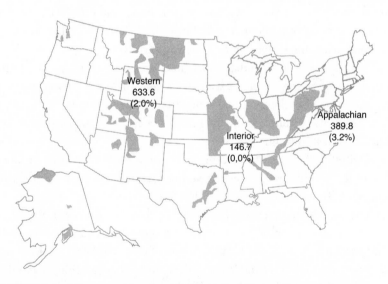

FIGURE 3.3 Where coal is mined in the United States. Percentages are changes in production from 2005 to 2006. U.S. total production is approximately 1,170 million short tons. *(Source: Energy Information Administration, Quarterly Coal Report, October-December 2008, DOE/EIA-0121)*[8]

How much coal does the world use?

Coal was important to the beginning of the Industrial Revolution, providing abundant energy in the 19th century but also causing great human
misery and environmental degradation. Because we don't use it much
anymore to heat our homes, and given that the picturesque coal-fired
steam engines are long gone, you may think people are using less and less
coal. But although it may not be very visible, worldwide use of coal is
actually growing faster than the use of the other fossil fuels, increasing
9% between 2006 and 2007, and 92% in the last 25 years.[9] The amount
of coal mined in the United States is also increasing, reaching record
levels in each year of the 21st century.[10]

People of the world currently use 3.4 billion tons of coal a year,
which provides 25% of the world's total energy use and 40% of the electricity. Coal is the major fuel used to produce electricity. (Petroleum, you
recall, leads in transportation and natural gas in home heating and manufacturing.) The United States gets about half of its electricity from coal,
and some nations depend even more heavily on coal for electricity production. For example, Poland and South Africa get 93% of their electricity from coal, China 78%, Israel 71%, and India 69%.

The United States, India, and China are expected to be the dominant users of energy in the next decades, and since these three depend
heavily on coal for electricity, it's going to be difficult to wean them off it.

Coal is also essential to making steel, because it is the carbon in coal
that turns iron into much harder steel. But only 2% of the coal used in
the United States goes into steel production; 92% is used to produce
electricity, and almost all the remaining 6% is used in other industrial
processes. Today Americans use only a small fraction—one-tenth of one
percent—for home heating. This is a big change from a century or so
ago, when many homes in the United States (and in Europe as well)
were heated with coal.[11] But the dirtiness of coal—its heavy, toxic smoke
spewing from chimneys and leaving black dust over everything, as well as
large quantities of ash that had to be shoveled out and disposed of—led
people to speedily abandon it for home heating as soon as oil and then
natural gas became readily available.

Are we going to run out of coal?

Eventually, yes, but since there's more of it, coal won't disappear as fast
as oil and gas will. At *current rates of use,* coal should last between

150 and 300 years.[12] But the key question is not the total amount in the ground, but the amount that can be recovered economically and energetically. As we mentioned earlier, *economically recoverable* means that a mining company can sell the coal it obtains for more than it cost to mine it. *Energetically recoverable* means that the amount of energy in coal at the point of use is greater than the energy expended to obtain it and transport it to that location, including the energy costs of all indirect activities, such as pollution control resulting from mining, other production costs, and transportation.

Mining coal

Mining coal is one of the major ways that people move the surface of the Earth. Of the 8 billion tons of stone, rock, gravel, sand, and soil that people move each year in the United States, 15% (1.1 billion tons) is from mining coal. That's almost as much as from farming (16%) and about one-third the amount that U.S. rivers move naturally (not counting the amount of material rivers remove because of erosion from agriculture).[13] Much of the rest (3 billion tons, almost half of all the earth-moving that Americans do) is due to road building.[14]

Strip mining

There are two kinds of coal mining: *strip mining* (also called *surface mining*) and *underground mining*. Strip mining provides more than half the coal in the United States and has been especially damaging to land, rivers, and ecosystems.[15] In the Appalachians alone, coal strip mining has destroyed more than 1,500 square miles—an area as big as Rhode Island—including 500 mountains and 1,000 miles of rivers and streams.[16]

A classic book about the effects of coal mining in the Appalachian Mountains is *Night Comes to the Cumberlands,* by Harry Caudill.[17] He tells the story of the poor Appalachian Kentucky farmers who owned the land in the Appalachian hills and mountains but were often unaware that they owned only the surface rights, not the mineral rights. (In the United States, rarely if ever does land ownership include owning the mineral rights. I learned this the hard way when I bought a house in Santa Barbara, once a major oil field, and attempted to buy the mineral rights. Try as I might, I was not able to purchase them, nor even to learn who actually owned them.)

Strip mining came to the Cumberland Mountains big-time in the 1950s. Previously, this mining method had been limited to comparatively flat lands to the west. But with the end of World War II, big machines— bulldozers and other earth movers—made it possible to strip the mountains.

The coal mining companies let the mountain landowners pay property taxes until the companies were ready to mine the coal. Then they would arrive with their big machines and knock down whatever was in the way, including houses. Legally, the mining companies "…could build roads wherever they desired, even through lawns and fertile vegetable gardens," Caudill writes. "They could sluice poisonous water from the pits onto crop lands. With impunity they could hurl out from their washeries clouds of coal grit which settled on fields of corn, alfalfa and clover and rendered them worthless as fodder. Fumes from burning slate dumps peeled paint from houses, but the companies were absolved from damages."[18] The story Caudill tells is one of the saddest in American environmental history. He writes:

> The cumulative effects of the wrecking of a coal-filled mountain stagger the imagination. Let us suppose the ridge contains three seams of coal, and that the company first strip-mines the bottommost seam. A few years later it returns to a higher seam midway up the mountain and cuts high walls of fifty or sixty feet in its sides. Then to crown its enterprise its shovels and bulldozers slice off the top of the mountain to recover all of the highest seam. Within a dozen years it has dug millions of tons of coal and made a profit of millions of dollars. But in the process it has totally transformed one of earth's terrain features. A relatively stable mountain, whose soil and water were to a high degree protected by grass and trees, has been reduced to a colossal rubble heap.[19]

As Caudill makes clear, strip mining completed the deforestation that had started with small-time logging in the late 19th century. It also removed the soil, thereby increasing sediment transport downstream. Sulfur in the coal—an impurity—acidified water draining out of a mine, denuding the land and destroying wildlife and fish populations, habitats, and ecosystems downstream.

The most damaging variation on strip mining is mountaintop removal (Figure 3.4): The tops of mountains are cut away, the coal is removed, and the waste is pushed into the valleys. This is obviously highly destructive, increasing the chance of floods and bringing toxic mine waste to the surface.

FIGURE 3.4 Mountaintop removal at the Martiki mine in Martin County, Kentucky. One such strip-mined area covers 7,000 acres. *(Photo by Vivian Stockman / www.ohvec.org. Flyover courtesy SouthWings.org.)*

In response to complaints, the federal government began to require that strip-mined land be restored and that mining companies build holding ponds to contain the acidified water. The results have been mixed at best. Relatively little land has been restored, and even what has been "restored" is little like what had been there before. Holding ponds fail— for example, an earth dam near Inez, Kentucky, gave way on October 11, 2000, releasing 250 million gallons of wastewater from coal mines, a spill larger than that of the *Exxon Valdez*. The release is said to have turned the river "an iridescent black" for 75 miles downstream and killed all aquatic life, including several hundred thousand fish, for more than 100 miles.[20] Kentucky has 58 such dams.

Surface-mining practices are supposed to have improved since the 1950s, and land is supposed to be restored. But a 2005 article in the Louisville, Kentucky, *Courier Journal*, written by a person who revisited old mining areas, said that the laws passed to "control the devastation of strip mining for coal . . . have been ineffective in controlling an industry that has gone wild since the energy shortage has driven the price of coal up to $100 a ton."[21] Other stories about strip mining include a description of one mine in Wyoming, abandoned about a half century ago and "restored," that still has no vegetation.[22]

Underground mining

We have focused so far on surface mining, but 40% of coal mining in the United States is underground, and of the 82,000 people who work in coal mining in the U.S., 53% are involved in underground mining versus 47% in surface mining.

Coal has been dug out of the ground and used as a fuel by people for thousands of years. In fact, some archaeological evidence suggests that coal was sought and used for heating perhaps more than 100,000 years ago. Coal mining appears to have been active in China for thousands of years, perhaps 10,000. In the Americas, the Aztecs were the first to use coal, both decoratively and as a fuel. European settlers began to mine coal in the mid-18th century. With the beginning of the Industrial Revolution, and especially with James Watt's invention of the steam engine, coal mining became a major activity, important for fueling the new industrial age.

At first, coal was obtained from places where it came to the surface or from shallow mines. Large-scale underground mining developed in the 19th century, and disasters—cave-ins and explosions—became part of it. The biggest coal mining disaster in Great Britain killed more than 400 miners. In the U.S., the worst was an explosion in a coal mine in Monongah, West Virginia, that killed 362 people. The collapse of part of a coal mine in Huntington, Utah, in August 2007, reminded people of the dangers of this work.

As tragic as these are, the larger-scale impacts on health and environment are worse. Underground mining, like surface mining, produces acid mine drainage. Sometimes land collapses above mines (this is called *subsidence*), and sometimes underground fires start in coal beds and persist for years over large areas. One of the most notorious started in 1961 in Centralia, Pennsylvania, with a trash fire that spread to coal seams and is still burning today (Figure 3.5). As a result, Centralia has become a ghost town.[23]

FIGURE 3.5   A fire that started in 1961 in an underground coal bed in Centralia, Pennsylvania, still burns in this photograph taken in January 2008. (PA Department of Environmental Protection. *Photos and commentary by Donald Davis, www.offroaders.com/album/centralia/photos36.htm*)

Centralia's fire isn't the only one. An underground fire that started around 1915 on land owned by the Red Ash Coal Company in Laurel Run, Pennsylvania, has continued over the years. In the 1960s, homes, schools, and stores were abandoned, and the fire is still burning today.[24] A total of 45 mine fires continue to burn in Pennsylvania.[25]

Underground coal fires are also a problem in China,[26] where such fires are estimated to release more greenhouse gases every year than all the cars in Germany, according to the International Institute for Geoinformation, Science and Earth Observation (ITC), in the Netherlands.[27] Worldwide, according to Andrew Revkin of the *New York Times,* "Fires are burning in thousands of underground coal seams from Pennsylvania to Mongolia, releasing toxic gases, adding millions of tons of heat-trapping carbon dioxide to the atmosphere, and baking the earth until vegetation shrivels and the land sinks."[28] The federal Office of Surface Mining says that the costs of such fires in the United States have reached nearly $1 billion.[29]

A large-scale form of underground mining is *longwall mining,* in which a long wall—up to about 1,200 feet long—is mined as a single unit in one long slice by a machine called a *shearer.* The shearer has cutting drums and hydraulic rams run by electric motors and is somewhat analogous to the huge machines that dig tunnels. Longwall mining is efficient and safer for miners, but it can lead to serious subsidence of the land surface after the walls of coal have been removed.

Burning coal as fuel is a major source of air pollution

Coal is one of the leading causes of smog, acid rain, other toxic substances that get into the air, and greenhouse gases. According to the Natural Resources Defense Council, "10.3 million tons of sulfur dioxide were released from U.S. power plants in 2004, 95 percent of these emissions coming from coal-fired plants." Similarly, more than 90% of the 3.9 million tons of nitrogen oxides released each year come from coal-fired units. Coal-fired power plants also put into the air approximately 48 tons of mercury, 56 tons of arsenic, 62 tons of lead, 62 tons of chromium compounds, 23,000 tons of hydrogen fluoride, and 134,000 tons of hydrochloric acid each year.[30] Coal produces one-third of all carbon dioxide emitted by human activities in the United States.

As most of us are aware by now, air pollution is a significant cause of death. An analysis by the Sierra Club suggests that in the U.S. alone "cutting power plant emissions by 75% could avoid more than 18,000 of the deaths caused by particle pollution."[31] In China, almost 14% of adult deaths result from pulmonary diseases, making these the second-largest single cause of adult deaths.[32]Although it is not easy to tie down the relationship precisely, pulmonary diseases appear to be closely related to air pollution and, therefore, to coal burning.[33]

Sulfur and nitrogen oxides emitted from burning coal create acids when mixed with water. This acid rain has been shown to affect bodies of freshwater and cause a variety of problems for freshwater life, including developmental malformations of freshwater animals. It is also believed to affect forest and grassland ecosystems, but the evidence is still mixed and not conclusive.

Black Mesa

One of the most notorious and contentious cases of air pollution from a coal-fired power plant in the United States is the Black Mesa project at the Four Corners (where Arizona, Colorado, New Mexico, and Utah meet) in the U.S. Southwest. Black Mesa mine is three interconnected stories: one about a huge coal strip-mining operation; another about the use of water from the Navajo aquifer to move the mined coal; and the third about the Mojave Power Plant that burned the coal and operated from 1971 until the end of 2005.[34]

The Black Mesa stories began in 1964 when the Peabody Western Coal Company signed a contract with the Navajo and Hopi tribes allowing the company to strip-mine coal on the Indian lands and pump water from the aquifer on the reservations to transport the coal to a huge new electric power plant. The Black Mesa mine, together with the Kayenta mine nearby, both operated by Peabody Western Coal Company, covered 62,753 acres on Navajo and Hopi land. The original agreement is contentious, with representatives of the two tribes claiming that a few individuals sold out to their own benefit without adequately reviewing the arrangement with the rest of the people of the tribes. The agreement allowed the Peabody Coal Company to remove 670 million tons of an estimated 21 billion tons of coal that lie within the area.[35] When operating, the mine produced 4.8 million tons of coal a year. And every day, 3.3 million gallons of groundwater was pumped from the Navajo Aquifer 120 miles away to create a slurry with the coal that could then be pumped to the power station (the only such coal-slurry transportation in operation).

The Navajo Aquifer lies beneath both Hopi and Navajo nations in northeastern Arizona and is considered by them sacred water. Now, however, the water stored in the aquifer has decreased, creating water-supply problems for the Navajo and Hopi. Black Mesa Indigenous Support, a group that describes itself as "350 Dineh residents of Black Mesa," states that "the Peabody Coal Company pumped 1.3 billion gallons of pristine water a year out of an ancient sandstone aquifer that lies beneath the Hopi and Navajo lands." This organization also states that, as a result, "wells and springs have dried up and the entire ecology of Black Mesa has changed. Plants have failed to reseed and certain native vegetation has died out." Water levels have dropped by more than 100 feet in some wells, the group says, and concludes that "these developments threaten the viability of the region's primary water source."[36]

Black Mesa coal fueled the 1,500-megawatt Mojave Power Station operated by Southern California Edison (SCE) and primarily owned by that company. SCE received 56% (885 megawatts) of the power, sharing the rest with three other owners: Salt River Project, which received 20% (316 megawatts); Nevada Power Company, 14% (221 megawatts); and the Los Angeles Department of Water and Power, 10% (158 megawatts).[37] It produced enough electricity for 1 million people in California.

Controversies about Black Mesa erupted both locally and nationally. Locally, the Navajo and Hopi tribes battled against the use of their water and destruction and pollution of their lands. Nationally, environmental groups battled against large-scale environmental pollution from the Mojave Generating Station, which emitted more than 40,000 tons of sulfur dioxide a year (making it one of the largest sources of this pollutant in the western states), as well as 19,201 tons of nitrogen oxides, 1,924 tons of particulate matter, and about 10 million tons of carbon dioxide.[38]

In 1998, three environmental organizations—Grand Canyon Trust, Sierra Club, and the National Parks and Conservation Association—filed suit against the owners of Mojave because of widespread pollution from the plant, reaching to the Grand Canyon and affecting visibility there since the 1980s.[39] The Navajo and Hopi tribes also sued the owners of the power plant several times.[40] One result was that the EPA established guidelines to reduce the Mojave Station's air pollutant emissions. It was estimated that the required pollution controls would cost $1 billion.[41]

The plant closed at the end of 2005 because of noncompliance, and it is unclear whether it will reopen.[42, 43]

Strip mining in the Kentucky Cumberland Mountains and strip mining and coal burning to produce electric power near the border between Arizona and Nevada are similar stories. Both involve large corporations and the people living on land above the coal. The seemingly easy money paid to residents to give up their rights and allow mining on their land destroys the land and its resources. Although a lot is said about making coal into a clean fuel, good for people, that has not been the general history of coal mining. Money from this "black gold" attracts people and corporations. In the past, including the recent past, laws have not been adequate to prevent large-scale environmental damage.

Financial costs of environmental damage from coal

It's hard to assess the total cost of mitigating environmental damage from mining and burning coal. Under U.S. federal law, funds are available to restore abandoned coal mine lands, as follows:

> Title IV of the Surface Mining Law—the Abandoned Mine Land Reclamation Program—provides for the restoration of lands mined and abandoned or left inadequately restored before August 3, 1977. Implementation is accomplished through an Emergency Program (for problems having a sudden danger that presents a high probability of substantial harm to the health, safety, or general welfare of people before the danger can be abated under normal program operating procedures) and a nonemergency program. States and tribes with approved programs carry out these responsibilities. Since 1979, when the states began receiving abandoned mine land administrative grants to operate their programs and construction grants to complete reclamation projects, through September 30, 2004, $3,579,356,901 was distributed from the fund. [44]

Since 1996, the Federal Surface Mining Law has authorized the government to award grants to regulate coal mining. By 2004, more than $3.5 billion had been paid out.[45]

The future of coal power

One good thing about coal is that there's a lot of it. Another good thing about coal is that it is cheap, at least with today's price supports and sub- sidies, and as long as only the extraction and transportation costs are included; energy from gas and oil costs about 6 times what it costs from coal. But in 2009 the U.S. Energy Information Agency recalculated the costs to mine coal and determined that at $10.50/ton, the cost a few years ago, only 6% of the coal in Wyoming, the country's largest reserve, would be economically recoverable.[46] At the time of this writing, the average sale price for Wyoming coal is $11.39 per ton, and the average price of all coal in the U.S. is $43.53/ton.[47] The present cost in the U.S. is $8.5/ ton. The question is at what cost per ton would the cost of electrical energy from coal equal that produced by wind and solar energy.

Installation costs for coal-fired power plants have traditionally been estimated at $1.50–2.00 per watt. But the cost is rising rapidly; one cur- rent report says that in 2010 the costs are on the order of $3.50 per watt.[48] At present, wind energy costs between $1 and $3 per installed watt. Let's consider the less expensive case, where the costs are $1 per watt and the average power output from an installed watt is 2.347 kilo- watt-hour over a year (a standard value). In the first year of use, the wind turbine would produce electricity at a cost of $0.43 per kilowatt-hour. Assume there are no maintenance or other costs, and therefore the total costs are contained in the installation, the cost after 10 years would be $0.043 per kilowatt-hour. At this cost per kilowatt-hour, wind matches electrical energy produced from coal when coal is $8.12/ton. Making the same assumptions for the wind turbine, after 20 years the cost would be $0.002 per kilowatt-hour, and coal would have to be priced at $0.41/ton to match this.

Solar is considerably more expensive, both because it costs more to install and because the yield per installed watt capacity is less. (Wind can blow at night, but the solar cells produce electricity only during the day.) Solar costs are typically estimated to be between $3 and $5 per installed watt. Let's consider the less expensive case, where the costs are $3 per watt and the average power output is 1.245 kilowatt-hours per watt installed (a standard value). Assuming that there are no maintenance or other costs, and therefore the total costs are contained in the installation, the cost after 10 years would be $0.24 per kilowatt-hour. At this cost per kilowatt-hour, solar matches coal when coal is $462.7/ton. Making the same assumptions for the solar devices, after 20 years the cost would be $0.012 per kilowatt-hour, and coal would have to be priced at $46.27/ton to match this.

In short, wind is already cost-competitive against coal, and solar is approaching being cost-competitive against coal, with installation costs averaged over 20 years. And this does not include direct pollution and land-conservation effects (strip mining, erosion, sedimentation, and so on) are taken into account.

But worldwide, coal is selling at much higher prices, and prices have been rising. We will explain this in more detail in Chapter 13.

Wide recognition of the problems with coal resulted in a decline in building new coal power plants in the U.S. However, since there is so much coal, and since the technology for using it has been around for centuries, and since major power companies have a lot to gain from selling coal and the energy from it, there is now and will be more economic pressure to use coal.

In 2006 there were 476 coal-fired electric power plants in the United States, and they produced 2 trillion kilowatt-hours.[49] Advocates for coal-fired power plants are working to have 150 more added.[50] More than 100 conventional coal-fired power plants are in various stages of development in the U.S. By 2007, 28 were under construction, 6 were near the construction stage, and 13 more had received permits. When these are completed, they will increase the number of coal-fired power plants by 10%. The Department of Energy projects that by 2030 the equivalent of 450 new large (300 MW) coal-fired power plants will be completed."[51]

China, too, is adding new coal-fired electric power plants. Right now, 78% of China's electrical energy is produced from coal. The Natural Resources Defense Council estimates that China's electrical power generation will increase from 600 billion watts in 2006 to 800 billion watts in 2010, much of it from coal. "China's coal sector is not only the world's largest, but also the most dangerous and most polluting," the NRDC added.

Who is promoting this, and why?

Of course, the coal mining corporations and electric power companies support continued use of coal, as do the U.S. government and other national governments. The Energy Policy Act of 2005 included $1.65 billion in tax incentives for new coal plants, $1 billion of which has been allocated to nine projects around the country.[52] World coal prices have risen (Figure 3.6), making it more attractive to mine, transport, and sell.

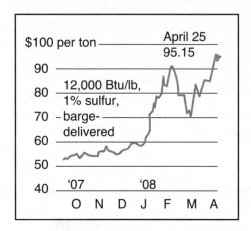

FIGURE 3.6 The price of coal rose sharply, doubling between 2007 and 2008 due to rising demand around the world. *(Source: AP Images/Platts, AP)*

In addition, some academics envision the continued use of coal. For example, a recent authoritative report by 13 distinguished scientists and scholars at MIT states that "coal will continue to be used to meet the world's energy needs in significant quantities." They go on to say that "CO_2 capture and sequestration (CCS) is the critical enabling technology that would reduce CO_2 emissions significantly while also allowing coal to meet the world's pressing energy needs."[53] They concede that "no CO_2 storage project that is currently operating…has the necessary modeling, monitoring, and verification (MMV) capability to resolve outstanding technical issues, at scale." However, they "…have confidence that large-scale CO_2 injection projects can be operated safely."

This report is one of many that views greenhouse gases as the main concern about coal, and the writers are optimistic that new technology can take care of this. But given the history of the use of coal and the magnitude of its environmental effects, is such faith in an unproven technology justified and a wise risk for the future, in comparison to other energy opportunities?

Technologies to make coal cleaner

Earlier in this chapter I wrote about FutureGen and the possibility that coal might become a "clean" fuel. The technology involved burying (sequestering) the carbon dioxide produced when coal burned. How feasible is this? It's not easy to bury carbon dioxide that comes off a hot coal fire. It's much easier to get rid of it the old way, by releasing it into the air from tall smokestacks. In June 2009, Southern Company and American Electric Power withdrew from the project, while that same month, the Obama administration put $1 billion into the project as part of the federal stimulus package.[54, 55]

Craig Canine, writing for the Natural Resources Defense Council, describes his visit to one of the experimental carbon-sequestering plants near Weyburn, Saskatchewan, built by PanCanadian Petroleum. The plant was using the CO_2 to force oil and gas out of deep wells—4,600 feet down—while at the same time burying the CO_2.[56] Basin Electric, the company generating electricity from coal, had to pump the CO_2 some 205 miles from the power plant to the oil field and had to pressurize it to 2,000 pounds per square inch to do so, requiring one of the most powerful compressors ever made, and powering these with two 20,000-horsepower electric motors. The plan is to bury 20 million tons of CO_2. Doing this for 20 years is expected to yield an additional 120 million barrels of oil from the field.

The Saskatchewan project is one of about a dozen around the world trying to bury CO_2. The recent report by MIT scholars also focused on technologies to bury carbon dioxide emitted by coal fires, but there are three other approaches to trying to make coal "burn cleaner." The first is to burn coal as it comes from the ground but use physical and chemical scrubbers to remove some of the pollutants on their way up the smokestack and neutralize acid-causing chemicals by adding limestone to the hot emissions. The second is to turn coal into a gas and then burn that. And the third is to turn coal into a liquid—basically into gasoline or diesel—and burn that. The first is self-explanatory, so let's start by taking a look at the second.

Turning rock into a gas and then burning it

GreenPoint Corporation, the brainchild of Andrew Perlman, is testing new ways to convert solid coal into a gas that will burn cleaner than the original. As you already know, this is not a new idea—remember the 19th-century gas streetlamp in the chapter's opening photo. What is new at the GreenPoint facility is a secret catalyst that is supposed to convert the solid coal to methane at temperatures much lower than previously needed and separate the pollutants from the gas. The pollutants still have to be disposed of somehow, but they are kept in one place rather than released into the atmosphere. The plant is experimental and has yet to operate.[57]

Turning rock into a liquid and then burning it

The third method proposed to make coal a "clean" fuel is to make a liquid fuel from it. In the 1920s, two Germans, Franz Fischer and Hans Tropsch, developed a way to do this. Their method, which was used by the Nazi government during World War II to provide fuel for military vehicles,[58] uses a catalyst that is supposed to leave the fuel much cleaner, with fewer particulate chemical pollutants.

Although making liquid fuel and some forms of gas fuel from coal are proven technologies, they add costs to the production of electricity and do not themselves dispose of toxic chemicals, dust, or ash. Therefore, it's unclear whether they will provide a net benefit environmentally, economically, or to society. At the time of this writing, the future of clean-coal technology seems uncertain but also seems to depend on heavily government subsidies. Funds are going into these technologies and will likely continue to do so. FutureGen, directed by the Department of Energy, is funded to the tune of $1 billion. Is this the best use of federal research dollars in a search for cleaner and secure energy sources for the future?

The bottom line

- Coal is the most abundant fossil fuel and, although not evenly distributed around the Earth, occurs in more nations than do large deposits of oil and natural gas.
- Coal remains a cheap fuel to buy, largely because of government subsidies and other benefits, and especially if the price doesn't include the costs of polluting the environment and damaging human health.

- Because of its availability and relatively low price, we can be fairly certain that coal use will increase in the next decades, especially to generate electricity.

- However, with few exceptions, the mining of coal, an ancient practice, has taken place at the cost of human lives, damage to health, and sometimes destruction of farms, towns and larger settlements, as well as natural environments—forests, wildlife habitats.

- The burning of coal is a major source of air pollution from soot and toxic chemicals that have affected human health, natural ecosystems, agricultural lands, wildlife, and freshwater fish. Coal dust also decreases visibility in the atmosphere.

- Burning coal is one of the major ways that people are adding carbon dioxide to the atmosphere. Hence, much of the recent concern about coal is about its effects on climate, and much of the current emphasis on "clean coal" is to reduce CO_2 emissions into the atmosphere.

- And finally, mining and burning coal mar the beauty of the land and its diversity of life.

4

Water power

Roll on, Columbia, roll on,
Roll on, Columbia, roll on,
Your power is turning our darkness to dawn.
So roll on, Columbia, roll on.
—Woody Guthrie

FIGURE 4.1 Edwards Dam on the Kennebec River, at Augusta, ME, just
before it was breached. *(© AP Images)*[1]

Key facts

- Water power provides about 10% of the electricity in the United States, at least half of the electricity used in about one-third of the countries of the world, and more than 90% of the electricity in 24 countries, including Brazil and Norway. Large dams generate 19% of the world's electricity.

- The U.S. has about 80,000 dams more than 6-feet tall, many built to provide electricity.[2] Worldwide, about 800,000 dams exist, and about 45,000 are "large," meaning more than 15 meters (about 45 feet) high.[3]

- The United States has 2,400 dams that generate electricity.

  About 17% of the rivers in the United States have dams, which in total block about 600,000 river miles.

- China has the greatest number of large dams, 22,000 of them.

- In developed nations, most or all the best sites for hydroelectric power have already been tapped, and there is little potential to increase this source of energy.

- Most of the remaining undeveloped water-power sites are in developing nations.

A story about water power: the breaching of Edwards Dam

In August 1999, I traveled to Augusta, Maine, to watch the breaching of Edwards Dam. I wrote this article about it, which appeared in the *Los Angeles Times* on August 22 of that year. It still seems to sum up my experiences, so here is what I wrote that day.[4]

> As a crowd gathered along the east bluff of Edwards Dam, a great blue heron flew low above the Kennebec River and was soon out of view, traveling downstream from where water still flowed smoothly over the 161-year-old structure. The heron had been disturbed from its usual stalking territory, perhaps by the big diesel shovel digging bucketfuls of soil from a temporary dam across the river, or perhaps by the large crowd on the opposite shore, or the noise of a helicopter and a float plane circling overhead carrying television crews.

They had all shown up today to witness the breaching of a major hydropower dam, an unprecedented event in U.S. history. The dam was being removed to save migrating fish, restore Kennebec's ecological habitats, and improve recreational fishing and boating. If the river's fish population increased, it might be a boon for the heron as well.

Built in 1837, the dam was operating when Henry David Thoreau canoed Maine's rivers in the 1840s. The dam's demise was "a bittersweet event," Augusta's mayor said. It was a willing and willful removal of one of the triumphs of the machine age, a piece of Yankee ingenuity that had provided power, jobs, and prosperity for Augusta, but did so no more.

Each century and each generation has had its own approach to water power and rivers. Dams were built throughout the United States to provide power for many kinds of industries, from textile mills to aluminum refineries; to store water for irrigation; to control water levels as an aid to ships transporting grain and other goods; for flood control; and for recreation. But dams greatly altered stream habitat. As Thoreau observed on the Merrimack, migrating fish such as shad and salmon, once common there, were rare in his time since they could not pass over the dams already in place by the 1840s.

When a big bell in an Augusta church announced the time to breach Edwards Dam, the shovel dug deeper and deeper into the earthen dam until a shout went up from the crowd and water began to spill through. The thin trickle, moving ever faster, turned into a frothy, mud-laden torrent running down the far side of the Kennebec, tumbling against an old mill building and turning the main channel brown as the river began to cleanse itself of 150 years of deposits behind the old dam.

Some would say that removal of the dam was righting an old wrong. Others would say it was a mistake, wronging an old right. But it was not a matter of absolute right or wrong, it was a matter of a change in our society's needs and desires, a continuation of change and progress, of new ideas, that have characterized U.S. society.

Edwards Dam was among the first of many to be removed. It was a relatively easy decision since the dam was old and dangerous and had ceased to be of much economic use to Augusta. Maine had begun to prosper more from tourism than from the dam and its mills. The industries along Augusta's riverbanks were closed or were using electricity generated far away, mostly from fossil fuels whose "brown" pollution didn't seem to touch Maine's scenic landscapes and rivers. The new environmentalism would save the city and its landowners a bundle, paying to get rid of something that might cost them plenty if it broke on its own and flooded buildings downstream. Yes, the dam seemed only a negative, threatening migrating fish such as shad, which were in trouble along the coast. The majority of people wanted the river back as a renewable resource for living things. Other sources for energy could be found.

Elsewhere, other dams have caused much more conflict, with both sides claiming the environmental high ground. Opponents of hydropower side with those who argued for the removal of Edwards Dam. Supporters argue that dams are an important way to limit greenhouse gas emissions, that the reservoirs favor many kinds of wildlife and are major sources of recreation.

So today, water power seems to be a Dr. Jekyll and Mr. Hyde. It's a Dr. Jekyll for global warming since it emits no greenhouse gases. It's a Mr. Hyde about river habitats and the biological diversity of life in rivers, and the living things that depend on life in the rivers, including many people. What's the right balance about water power? Can it be an important source of energy in the future?

How much of our energy supply comes from water power today?

At present, the total U.S. hydroelectric power capacity, including pumped storage facilities, is about 95,000 megawatts, producing 2.14 million megawatt-hours[5]—about 10% of all the electricity and 3% of the total energy used in the United States.[6] The world's total electrical production from hydropower is about 2.3 trillion kilowatt-hours of electricity each year—about 19%[7] to 24%[8] of the world's electricity. In some nations, such as Canada and Norway, most of the electricity is produced by water power.

To generate all of its hydroelectric power (Figure 4.2), the United States uses 2,400 dams. (The rest of the nation's 80,000 dams don't generate electricity.)[9] About 17% of U.S. rivers have dams, estimated to block about 600,000 river miles. In the U.S. West, where water tends to be more limited than in the East, dams block almost every major river.[10]

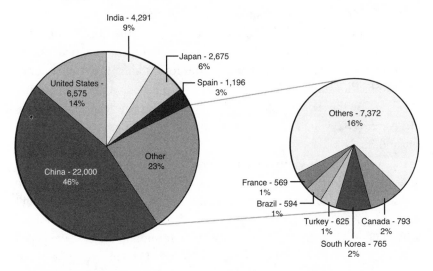

FIGURE 4.2 Percentage of the world's dams in major nations. *(Source: World Bank)*

The energy that dams provide, and how they do it

Water power is one of civilization's oldest sources of energy, familiar to us in paintings and photographs of water mills and their water wheels of medieval Europe and earlier. These used mechanical linkages—belts, gears, pulleys—to transfer the energy in the water to do useful work. But with the modern industrial age and the invention of the electric motor and electric generator, energy in flowing water generated electricity with much greater efficiency and flexibility. Soon hydroelectric power was doing all sorts of things, from providing electricity for the milkmaid to converting bauxite ore into aluminum (as Woody Guthrie put it in one of the songs he wrote for the Bonneville Power Administration).

The first big industrial-age improvement was the invention of a turbine wheel, a rotating fin-bladed device inside a metal housing, through which a large part, or all, of a stream's water could flow. These replaced the picturesque but less efficient water wheels that could make use of only a small part of a stream or river's energy. The first turbine wheels were linked mechanically to machines, just like their predecessors. The next big improvement was the invention of hydroelectric generators—huge, efficient machines spun by water flowing within them, just as air moves within modern jet engines, and generating electricity directly.

Then in response to the Great Depression and the Dust Bowl of the Midwest, Franklin D. Roosevelt's administration began construction of many major dams and created two entities for building them and operating them: the Tennessee Valley Authority (TVA), in 1933, which built 48 dams in the Southeast, and the Bonneville Power Administration (BPA), in 1937, which runs 12 major and 19 smaller dams on the Columbia and Snake rivers.

Later, the Army Corps of Engineers built six major dams on the Missouri River (mostly for flood control, irrigation, and recreation, but also for some electrical production), and the federal government also built some of our nation's largest and most impressive dams on the Colorado River: Glen Canyon Dam, with its huge reservoir, Lake Powell, just upstream from the Grand Canyon; and Hoover Dam and its Lake Mead reservoir, downstream from the Grand Canyon. This dam made it possible for Las Vegas to become a huge urban area.

Until the 1980s, dams seemed an unmixed blessing—water power was considered good for both society and the environment. In the days of the Depression, the Dust Bowl, and the construction of the first of these great dams, Woody Guthrie, a great American folksinger and songwriter, was hired by the Bonneville Power Administration to write songs promoting the dams. He wrote more than 20 songs, some still famous, all rejoicing in the social benefits of the dams. He wrote that "king salmon" and the great Douglas fir trees would have to give way to the dams and their reservoirs, but this was okay, he said, since it was more important to provide electricity "for the milkmaid." At a time when so many Americans were impoverished, building great hydroelectric dams seemed only a social good for all the people.

Downsides of major power dams, especially building new ones

I don't think I should have voted for the Glen Canyon Dam. Even though it's created the biggest tourist attraction in my state, I preferred the free-running river. I remember the river.

—Former U.S. Senator Barry Goldwater[11]

Nowadays, building big dams is controversial and time-consuming. A case in point is a large dam to be built on the Nam Theun River, a tributary of the Mekong River in Laos. The dam, named Nam Theun 2 (NT2), was proposed in the 1980s and became a major goal for the Laotian government in the early 1990s,[12] but work on it didn't begin until much later. By 2006, water diversion to build the dam had been completed, and people who lived on land that would be flooded were moved to new villages. Construction was finished in 2009 and on July 3, 2009, the hydro-generators sent out their first electricity, 60 megawatts.

NT2 is said to have cost $1.45 billion and at full operating capacity will produce 1,070 megawatts, 995 of which will be sold to Thailand, leaving 75 MW for use within Laos.[13] It is the largest single foreign investment ever made in that country, and support for it comes from the World Bank, the Asian Development Bank, the European Investment Bank, and the Nordic Investment Bank.[14] The World Bank estimates that the electricity sold to Thailand will give the Laotian government $30 million a year. The payback from Thailand should therefore take 50 years to equal the cost of building the dam,[15, 16] although there is hope that the revenue will increase substantially. The World Bank supports the dam since it fits the bank's goals of alleviating poverty and promoting economic development.

However, in 1997, in response to growing concerns about the ecological effects of major power dams, the World Bank and the Laotian government set up a panel of scientific experts to study the region's biodiversity, determine what the environmental and social effects would be, and recommend how best to deal with them. The three members of the panel were David McDowell, former director of the International Union for the Conservation of Nature and past New Zealand Ambassador to the United Nations; Ted Scudder, an anthropologist from Caltech and author of *The Future of Large Dams*;[17] and Lee Talbot, one of America's leading environmentalists and conservation scientists and primary author of the U.S. Endangered Species Act. This is clearly one of the most qualified panels that environmentalists and social scientists could hope for.[18] The idea was that this panel could lead a revolutionary effort to help create a dam that is benign and helpful to people and the environment.

Even with this distinguished panel, the project met with considerable criticism. The organization International Rivers states that the project did not fulfilled its promise to deal adequately with, first, "resettlement of 6,200 indigenous peoples on the Nakai Plateau; second, the program to mitigate NT2's impact on tens of thousands of downstream villagers; and third, compensation for villagers who have lost land and resources as a result of project construction."[19]

One of the displaced villagers said, "Before we depended on the land, now we depend on the company supporting us. We used to live next to the river and could get up in the morning and catch fish for breakfast."[20] It is estimated that the dam will directly affect 2% of Laotians.

A poor nation with fewer than 6 million people, Laos sees the dam as an important step forward. Although it will reduce logging in the nation's tropical rain forests, which has been a major economic activity, it will provide a variety of new jobs linked to the power that it will generate.

Three Gorges Dam

No discussion of the modern development of water power would be complete without considering the Three Gorges Dam, the largest in the world and the most famous built in recent years. The building of Three Gorges Dam on the Yangtze River in China created a worldwide controversy that fuels the debate about hydroelectric dams.

On the upside, Three Gorges Dam offers huge power output and no greenhouse gases. Three Gorges can produce about 18,000 megawatts of electricity, the same as 18 large coal-burning power plants. (In contrast, the largest hydroelectric dam in the United States, the Grand Coulee on the Columbia River, produces 6,180 megawatts, one-third of Three Gorges' output.) China's largest fossil-fuel resource is coal, and pollution from mining and burning coal is a serious problem in that country. Indeed, China and the United States are seen as the primary contributors to greenhouse-gas emissions. So the really big advantage of Three Gorges is that although it generates as much electricity as 18 coal-fired power plants, *it does not produce greenhouse gases.*

BUT—it has resulted in widespread flooding. This huge dam, about a mile long and about 600 feet high, created a vast reservoir 370 miles long, longer than the distance from Los Angeles to San Francisco, or from Boston to Philadelphia. It flooded many small farms and rural villages, directly displaced about 2 million people, and changed scenery that had been considered among the most famous and beautiful in the world.

Pollution and sediment are also big problems. Raw sewage and industrial pollutants flow into the Yangtze River and then into the reservoir. This river also has a naturally high sediment load, increased by human activities upstream and likely to increase further with China's rapid industrialization. As with all dams, river sediments fill in the reservoir, shortening the lifetime of the dam. Among other negative effects, these sediments will damage deepwater shipping harbors on the reservoir.[21]

Three Gorges Dam also lies within an area of active earthquakes, with their accompanying large landslides, raising international concern that a major quake could damage the dam, flooding the city of Wushan downstream, killing many of its several million people, and heavily damaging human development and the landscape downriver.

Some opponents of the dam—not opposed to water power in general but to Three Gorges specifically—point out that a series of smaller dams on the Yangtze River tributaries could have generated as much electric power with less environmental, scenic, and social damage, and without such serious danger from earthquakes.

Dams, in sum

What is the best answer to the question of whether dams should be an important source of energy in the future? Every new dam will cause some damage to wildlife habitat and biodiversity, in settled areas force the migration of many people, and in most cases cause the loss of some of the world's greatest scenery. The choice is between these undesirable effects and a dam's contribution to reducing greenhouse gas emissions and therefore the potential effects of global warming.

Perhaps the best summary of the role of large dams in our energy future was written by Thayer Scudder, author of *The Future of Large Dams*:

> Adverse environmental and social impacts of large dams are contributing to serious degradation of global life-support systems. Environmental impacts are degrading river basin ecosystems that, in many countries, are national heartlands. Social impacts are further impoverishing tens of millions of river basin residents who, to survive, cause further degradation of surrounding natural resources. Furthermore, the development potential of large dams is seldom realized since of the complexity involved, since of institutional inadequacies, since of implementation uncertainties and since of corruption.
>
> …Yet large dams remain a necessary development option for providing water and energy resources to populations in late-developing countries that are in crisis since they have expanded beyond the carrying capacity of their environment. Even then, a decision to build a large dam should only be based on an open and transparent options assessment process in which relevant stakeholders are fully informed of the risks involved. If a decision is made to proceed, state-of-the-art guidelines should be followed in order to reduce known disadvantages, some irreversible, to the extent possible and to increase an equitable distribution of advantages.[22]

Harnessing the power of rivers without dams

An alternative to damming a river is to use hydrokinetic devices— machines that float either on or below the surface of the river and generate electricity from the energy in the river's current. A surprising example is a plan to put a submerged floating turbine generator in New York City's East River. The turbine has been developed, but its installation has been delayed by concerns about effects on the river's fish. Verdant Power, the company that developed this turbine, says that the river's currents and width would allow enough of these turbines to generate as much as 10 megawatts of electricity in the East River without disrupting ship and boat traffic. The experimental machines are slated to power a Gristedes grocery store on Roosevelt Island.[23]

Can water power help solve the problem?

Could water power's contribution to our energy supply be increased greatly in the future? No, or at least highly unlikely. But if so, is it a good way to go? The answer is yes and no. As for conventional approaches to water power—large dams with electrical generating stations fixed in place—most good sites are already in use in North America and Western Europe. Good sites still undeveloped for water power exist elsewhere but are environmentally and societally controversial. Like the new dam in Laos and Three Gorges Dam in China, many remaining conventional water power sites are in developing nations and newly emerging industrializing nations. In both situations, new dams will affect biological diversity and displace many people, forcing them to seek new places to live and often new kinds of work. Other locations with still-abundant water power are wildernesses and remote places like eastern Siberia, where biological diversity will be affected. There is strong environmentalist opposition to any new dams.

Thinking more broadly, is there any way that even a combination of conventional and novel technologies could allow water power to significantly increase America's energy supply? The Electric Power Research Institute, a nonprofit corporation funded by the major electric power companies, estimates that enough locations are available, both tapped and untapped, to increase U.S. hydropower capacity 24% to 27%.[24] This would include increasing the generating capacity at existing hydroelectric dams; adding generators to dams that don't have them, building some small, low-power dams; using the new hydrokinetic technology to generate electricity from flowing water without building a dam; and also using ocean waves (see Chapter 8, "Ocean Power," on ocean energy). Excluding ocean energy would lower the projected energy increase to 14% to 15% and boost water power's contribution to total U.S. energy use from 3% today to about 4%.

Worldwide, the picture is brighter in terms of potential energy production (and ignoring, for the moment, social and environmental effects). At present, total world use of hydropower is 2,900 billion kilowatt-hours, about 2% of the total energy used by all the people of the world.[25, 26] According to the International Hydropower Association, an organization that promotes water power, hydropower could increase to about five times the present level, providing as much as 8% of today's worldwide energy use. However, not all this potential growth appears

economically feasible. About one-third (36%) of the world's economically feasible water power sites have been developed. "Most of the
remaining hydropower potential is in Africa, Asia, and Latin America."[27]
The International Hydropower Association observes that "two billion
people in developing countries have no reliable electricity supply, and
especially in these countries for the foreseeable future, hydropower
offers a renewable energy source on a realistic scale."[28] Proponents
argue that water power "continues to stand as one of the most viable
sources of new generation into the future."[29]

In sum, we conclude that hydropower can provide only a minor
overall contribution to the world's energy use, but locally and region-
ally—as in China, India, and other developing Asian nations and some
parts of Africa and South America—water power is almost certainly
going to increase unless other forms of alternative energy become easily
available.

The bottom line

- Water power is clean, doesn't pollute the air or soil, and doesn't
  contribute to global warming.

- However, hydroelectric dams and reservoirs alter river and stream
  habitats and are now targeted for removal by many environmental
  groups concerned about loss of habitats for fish, such as salmon,
  and streamside wildlife and vegetation.

- More than 650 dams have been removed in the United States and
  more than 200 since 1999.[30] Removals are in part to improve habitats, but also because some dams have become unsafe.

- Most good sites for hydropower are already in use in North America and Western Europe. Good sites yet undeveloped for water
  power exist elsewhere, but these are environmentally and societally controversial. They offer local and regional benefits at the
  price of undesirable local and global effects, including the dis-
  placement of people and destruction of their way of life.

5

Nuclear power

FIGURE 5.1 Indian Point Nuclear Power Plant stands 24 miles north of New York City's 8 million people. *(Photograph by Daniel B. Botkin)*

The Indian Point plant lies along the shore of the Hudson River estuary, famous since European settlement (and probably before written history among American Indians of eastern North America) for its scenery. The power plant's license is up for renewal, raising the question of whether nuclear reactors of this size should be so near major cities.

Key facts

- About 161 million Americans—more than half the population—live within 75 miles of one of the 104 nuclear power plants in the United States.

- The United States would need 1,000 new nuclear power plants of the same design and efficiency as existing nuclear plants to completely replace fossil fuels.

- Conventional nuclear power plants are not a short-term solution to the energy problem—they are complex and time-consuming to build and are controversial, hence even the siting of a plant takes time. From planning to going online would take a decade or more.

- More important, they're not a long-term solution either. The International Atomic Energy Agency, which promotes nuclear energy, says there are a total of just 4.7 million tons of "identified" conventional uranium stock that can be mined economically. If we switched from fossil fuels to nuclear today, that uranium would run out in four years. Even the most optimistic estimate of the quantity of uranium ore would last only 29 years.

- The life of a nuclear plant is just 30–40 years, and it costs more to dismantle it than to build it. Estimates for decommissioning and dismantling a large nuclear power plant run from $200 million to $500 million.

- Available federally sanctioned radioactive-waste disposal sites were said to be filled in 2008.

- According to U.S. government estimates, some 70,000 tons of highly radioactive nuclear waste are stored in temporary facilities. To move these across the country, such as to Yucca Mountain, Nevada, by truck and train would require one to six trainloads or truck convoys every day for 24 years. Now that this site has been rejected by the current administration, there is no planned permanent disposal site.

- The government believes the wastes will remain so toxic for 10,000 years that some kind of warning sign will be needed for that long.

Indian Point: the nuclear power plant in New York City's backyard

In 1974, over the vehement objections of Westchester County neighbors, the first of three nuclear reactors was built at Indian Point Power Plant in Buchanan, New York, 24 miles north of New York City (Figure 5.1). The power plant is 7.15 miles by road from the house I grew up in, and only a few miles away as the helicopter flies. Indian Point's second reactor was built two years later, and a third still later. The power plant—minus Unit 1, which was decommissioned in 1974—has been in operation since then, with a capacity of 2,000 megawatts. But Unit 2's license runs out in 2013, Unit 3's in 2015, and under U.S. law nuclear power plants must be relicensed. At the time of this writing, the controversy over the plant's relicensing continues, and whether to approve it has not been decided.

The National Regulatory Commission (NRC) announced the beginning of the process of relicensing the Indian Point Power Plant on May 2, 2007. By 2008, the relicensing of the plant had become a regional controversy, opposed by the New York State government, Westchester County, and a number of nongovernmental environmental organizations. The plant has operated for 22 years, so what's the problem?

Originally operated by Consolidated Edison and the New York Power Authority, over the years Indian Point had some difficulties. In 1980, Unit 2's building filled with water (an operator's mistake). In 1982, that unit's steam generator piping began to leak and radioactive water was released. In 1999, the unit shut down unexpectedly, but operators didn't realize it until the next day, when the batteries that automatically took over ran down.

Today, Entergy operates Indian Point. Under its management, a transformer burned in Unit 3 in April 2007; radioactive water leaked into groundwater, and the source of the leak was difficult to find. These are definite problems, but so far no catastrophic failure has occurred. Which is fortunate, since 20 million people live within 50 miles of this power plant. According to the *New York Times*, Joan Leary Matthews, a lawyer for the New York State Department of Environmental Conservation, said: "Whatever the chances of a failure at Indian Point, the consequences could be catastrophic in ways that are almost too horrific to contemplate."[1]

In addition to its scenic fame (Figure 5.2), the Hudson was made even more famous when folksinger Pete Seeger helped lead a cleanup of the river, one of the first major river restorations of modern environmentalism in the late 20th century. Before that, in spite of its well-known beauty, the Hudson, like all rivers, had been viewed since European settlement mainly as a means of transportation and a place to dump wastes. General Electric Corporation (GE) polluted the river with vast quantities of PCBs used in the manufacture of electrical equipment. Major lawsuits resulted, and ultimately a court ruled that the chemical was impossible to clean up directly and that GE should fund an organization that would help restore the river. This created the Hudson River Foundation and made one of America's most beautiful rivers the focus of intense restoration.

FIGURE 5.2 The Hudson River at Croton Point. Just a short way downriver from Indian Point, Croton Point has some of the Hudson's most beautiful scenery. *(Photograph by Daniel B. Botkin)*

Clearly, operating a nuclear power plant at this location not only presents a threat of disaster to 20 million people but is also inconsistent with the goal of preserving America's scenic beauty.

Nuclear power: no longer new, but suddenly popular

Not long ago, we would have put nuclear power in Section II of this book as a new alternative source of energy. Now we have to list it with our conventional energy sources because it became so important during the second half of the 20th century and is so widely used today. In fact, many people argue that nuclear power is a reasonable replacement for fossil fuels, a suggestion that has grown louder and more insistent with the growing concern about global warming. In 2006 a *New York Times* editorial endorsed nuclear power. The same year, the famous British environmentalist James Lovelock, whose Gaia hypothesis links all life to the global environment, also said we should turn to nuclear power. Stewart Brand, the originator and publisher of the *Whole Earth Catalogue* and whom the *New York Times* calls one of the originators of environmentalism, was quoted in that paper on February 27, 2007, as saying that he is for nuclear power and feels "guilty that he and his fellow environmentalists created so much fear of nuclear power." Even Patrick Moore, who claims to be one of the founders of Greenpeace, has become a spokesman for the nuclear power industry, according to the *Times*.[2]

Why are these well-known environmentalists in favor of nuclear power? Patrick Moore put the environmentalist argument succinctly. He wrote: "Wind and solar power have their place, but since they are intermittent and unpredictable they simply can't replace big baseload plants such as coal, nuclear and hydroelectric. Natural gas, a fossil fuel, is too expensive already, and its price is too volatile to risk building big baseload plants. Given that hydroelectric resources are built pretty much to capacity, nuclear is, by elimination, the only viable substitute for coal. It's that simple."[3]

Stewart Brand put it similarly. While acknowledging that nuclear power has its dangers and drawbacks, he has said that "it also has advantages besides the overwhelming one of being atmospherically clean. The industry is mature, with a half-century of experience and ever-improved engineering behind it. Problematic early reactors like the ones at Three Mile Island and Chernobyl can be supplanted by new, smaller-scale, meltdown-proof reactors like the ones that use the pebble-bed design. Nuclear power plants are very high yield, with low-cost fuel. Finally, they offer the best avenue to a 'hydrogen economy,' combining high energy and high heat in one place for optimal hydrogen generation."

Hugh Montefiore, former Anglican bishop of Birmingham, England, and for 20 years a Friends of the Earth trustee (who resigned over this issue), said that he is pronuclear because "the dangers of global warming are greater than any other facing the planet," and that "as a theologian, I believe that we have a duty to play our full part in safeguarding the future of our planet." He sees global warming as the holocaust, and therefore believes that "it is crucial if the world is to be saved from catastrophe that non-global-warming sources of energy should be increasingly available after 2010." He concludes: "I can see no practical way of meeting the world's needs without nuclear energy."[4]

In short, these three believe that global warming is by far the greatest threat to the planet, that no other form of energy is available in sufficient supply to replace fossil fuels, and that therefore, despite its dangers, it is necessary (Moore), and besides that, it isn't so dangerous anymore (Brand).

If leading environmentalists are for it, and the big power industry is for it—usually two opposing sides in the environmental debate—then this must be the way to go, right? Maybe not. But statements like these by Brand, Moore, and Montefiore, as well as the endorsement of nuclear power by such media institutions as the *New York Times*, were among the things that motivated me to begin a detailed examination of all energy sources. In particular, I wanted to find out whether, as Stewart Brand believed, nuclear had become safer, whether it could realistically be seen as a large-scale source of energy, and, most important, whether it truly was the only alternative to fossil fuels.

Nuclear energy today and tomorrow?

Today, nuclear power provides one-sixth of the world's electricity and 4.8% of the total energy. In the United States, 104 nuclear power plants produce about 20% of the country's electricity and about 8% of the total energy used (Figure 5.3).

As for tomorrow, here's the bottom line: The International Atomic Energy Agency (IAEA), which advocates and promotes the use of nuclear energy, states that the "total identified amount of conventional uranium stock" that can be mined economically is 4.7 million tons. According to my calculations, this means that if nuclear energy replaced all fossil fuels tomorrow, that quantity of uranium fuel would run out in

4 years. Even using the most optimistic estimate of uranium ore, it would last only 29 years.[5]

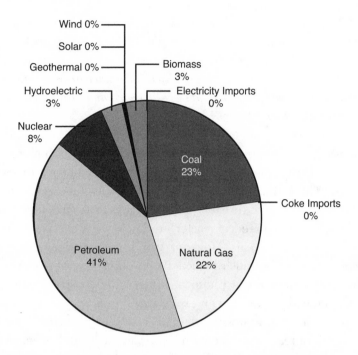

FIGURE 5.3 United States energy use (percentage by type) *(Source: U.S. DOE, EIA)*

Here's how I arrived at this conclusion. Today, nuclear energy consumes about 70,000 metric tons of uranium ore each year to provide 4.8% of the world's total energy use. Fossil fuels provide 87% of the world's energy. For conventional nuclear power plants to replace all fossil fuels, the energy obtained from those plants would have to increase 17.4 times, which means using 1.2 million tons of uranium ore each year. You just divide 4.7 million by 1.2 million.

The United States Geological Survey gives even more conservative estimates, stating that 3.3 million tons of uranium ore are available worldwide if the ore is priced at $130 per kilogram (the high end of present prices for this ore), and that there is an "inferred" amount—which I believe means the amount assumed but not determined to be out there—of 5.5. million tons. These estimates imply that the amount of fuel available would be used up in less than 3 to 5 years. *Thus, if the goal is to counter global warming by replacing all fossil fuels with nuclear power, this goal cannot be met.*

If the goal is to replace just petroleum and natural gas, because these are running out faster, then nuclear fuels would have to provide 63% of the world's energy, an increase of 13 times, which means annual use of 910,000 tons. *At that rate of use, the lifetime of nuclear ore and conventional nuclear power plants would be 5–38 years, and uranium ore would run out before either oil or gas.*[6]

Why isn't this common knowledge? Instead, the IAEA is quite optimistic about nuclear power's future, stating that "based on the 2004 nuclear electricity generation rate of demand the amount is sufficient for 85 years." This estimate assumes that 2004's nuclear energy production will continue into the future, but the IAEA goes on to state immediately that "fast reactor (breeder reactor) technology would lengthen this period to over 2,500 years. However, world uranium resources in total are considered to be much higher. Based on geological evidence and knowledge of uranium in phosphates the study considers more than 35 million tonnes is available for exploitation."

This leaves the impression that all is well for nuclear-reactor fuel. However, "fast reactors" (breeder reactors) are the kind that can be used to make fuel for atomic bombs. A few experimental breeder reactors were built by the U.S. government, but they were shut down or work on them halted in the 1990s. They are the kind of nuclear reactors that everybody fears Iran or North Korea might build and use to make atomic bombs. Other nations have tried building them, and some are considering or developing them, but to my knowledge no breeder reactor is being used to provide electric energy anywhere in the world. There are good reasons for this: The technology is not there yet, and the reactors are dangerous in themselves, even without considering their potential use in making atomic weapons.[7,8]

Somebody is sure to say, "But we've always found more oil, gas, and coal when we needed it. So can't we just wait for that to happen?" This is the Potato Creek Johnny gold prospector's approach. "It must be out there somewhere, we'll just keep moseying along until we stumble on something." You can take that approach if you want, but you will be ignoring the best-educated prospecting that has been done and is being done. It ignores the estimate by IAEA that allows for as much as seven times as much uranium ore as is economically available to be found out there somewhere. Indeed, some advocates of nuclear power say that we could concentrate dissolved uranium salts from the ocean and use that. Sure, and at what energy efficiency? My point is that if we want to plan the best we can, we cannot take this approach. It is the muddling through that has always gotten civilizations into trouble.

Today, the major nation that generates the greatest percentage of its electricity by nuclear power is France, with 78% (see Table 5.1). Belgium is second, with 60% of its energy from nuclear plants; Sweden is third with 43% from nuclear. Spain gets about one-third of its electrical energy from nuclear, which makes that nation an especially interesting one since, unlike the United States and many other developed nations, Spain does not get most of its energy from fossil fuels.

Table 5.1   Percent of Electricity Generated by Nuclear Power in Various Nations

Country	% Total	Generation (million kWh)
France	78%	368,188
Belgium	60%	41,927
Sweden	43%	61,395
Spain	36%	56,060
S. Korea	36%	58,138
Ukraine	33%	75,243
Germany	29%	153,476
Japan	28%	249,256
United Kingdom	28%	89,353
United States	19%	610,365
Canada	18%	94,823
Russia	12%	119,186
World Totals°	18%	2,167,515

° World totals include countries not individually listed.
Source: *Energy Studies Yearbook: 1993* (New York: United Nations, 1995).

It is also interesting to note that almost one-third of all nuclear power plants are in North America and another third are in Western Europe (Figure 5.4). In contrast, all of Asia, with most of the human population, has only about one-fifth of the world's nuclear energy production. Thus, right now nuclear power plants are largely an issue in North America and Western Europe—that is, in the first major nations to become industrialized and that now have among the world's greatest energy use and highest standards of living.

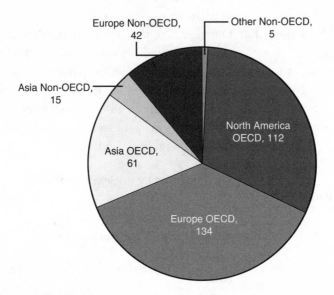

FIGURE 5.4   World nuclear generating capacity in 2004, in billions of watts, by geographic region, for the richest countries in the world (members of the Organization for Economic Cooperation and Development, or OECD).[9, 10] This shows that for the richest countries in the world, nuclear power plants and their problems are right at home.

Where exactly do you find uranium ore?

If the limited amount of uranium ore in the world is not enough of a problem, consider how these ores are distributed around the world (Table 5.2). Australia has the largest amount, 22.7%; Kazakhstan and Russia are second and third; and except for Canada and the United States itself, the other leading nations are not necessarily good sources for the United States. One can imagine declining supplies of uranium giving rise to the same global conflicts generated by dwindling oil and gas deposits.

Table 5.2   Nations with the Largest Uranium Ore Deposits

Rank	Nation	Tons	%
1	Australia	1,243,000	22.7%
2	Kazakhstan	817,300	14.9%
3	Russia	545,700	10.0%
4	South Africa	435,100	8.0%
5	Canada	423,200	7.7%
6	USA	339,000	6.2%
7	Brazil	278,400	5.1%
8	Namibia	275,000	5.0%
9	Niger	274,000	5.0%
10	Ukraine	199,500	3.6%
11	Jordan	111,800	2.0%
12	Uzbekistan	111,000	2.0%
13	India	72,900	1.3%
14	China	67,900	1.2%
15	Mongolia	62,000	1.1%
16	Denmark	32,300	0.6%

Source: *Uranium 2005: Resources, Production and Demand,* jointly prepared by the OECD Nuclear Energy Agency (NEA) and the International Atomic Energy Agency (IAEA).

Who pays and who benefits?

Nuclear power is expensive. According to the Nuclear Energy Information Service, which calls itself "Illinois' Nuclear Power Watchdog for 25 Years," nuclear power has cost $492 billion, "nearly twice the cost of the Viet Nam War and the Apollo Moon Missions combined."[11] The corporations that built nuclear power plants would be operating them at a loss (or trying to shut them down) if they weren't benefiting from heavy government subsidies, paid for by your tax dollars.

In spite of these limitations and problems, the Obama administration is moving ahead with the federal funding of nuclear power plants. At the time of this writing, the administration had allocated $18.5 billion for new "next generation" nuclear power plants, to be divided among UniStar Nuclear Energy, NRG Energy Inc., Scana Corp, and Southern Co.[12]

How safe are nuclear power plants?

Advocates of nuclear power argue that it is safer than other sources of energy. They say that the number of additional deaths caused by air pollution from burning fossil fuels is much greater than the number of lives lost through nuclear accidents—for example, the 4,000 deaths that can be directly attributed to the Chernobyl nuclear power accident and the forecast 16,000 to 39,000 deaths that might eventually be attributed to Chernobyl—are fewer than the number of deaths each year caused by burning coal.[13] Those arguing against nuclear power say that as long as people build nuclear power plants and manage them, there will be the possibility of accidents. We can build nuclear reactors that are safer, but people will continue to make mistakes, and accidents will continue to happen. And beyond the possibility of accidental disaster is the now all too real possibility of deliberate disaster.

Aren't nuclear power plants safer today?

Perhaps not. Here are some examples of recent problems with nuclear power plants. According to Florida's *Sun Sentinel* newspaper, videotapes at the Peach Bottom Atomic Power Station, a nuclear power plant in Pennsylvania (60 miles south of Harrisburg on the Susquehanna River), showed guards sleeping on the job in September 2007. The following year, on April 10, 2008, the federal Nuclear Regulatory Commission announced that it was going to fine Florida Power & Light $130,000 because six security guards at FPL's Turkey Point nuclear power plant in Homestead, Florida (just 35 miles south of the center of Miami, next to Biscayne National Park and near Florida's Seaquarium), were repeatedly caught sleeping on the job between 2004 and 2006. It seemed that one of the jokes from "The Simpsons" television program—Homer Simpson falling asleep at his job at a nuclear power plant—was coming true.[14] Earlier in 2008, the NRC had fined the utility $208,000 for failing to provide acceptable equipment to security employees.

While this was happening, FPL was proposing to build two new nuclear power plants, at a cost of $12–24 billion, at the same Turkey Point facility.

The problem of security guards sleeping on the job is familiar to anyone with military experience—one of the hardest things to maintain is alertness when nothing happens most of the time, even though there is always a chance of something very bad happening. As nuclear power plants begin to seem more and more ordinary, and as time passes without incident, people charged with monitoring them and protecting us will be lulled into letting down their guard.

Radioactive waste

I became acquainted firsthand with radioactive waste when I was a graduate student. It was the mid-1960s and the use of radioactive chemicals in scientific research was still pretty new. This was especially true in ecology, where radioactive elements were beginning to be used to trace chemicals in the environment; to study nutrition of animals and plants, food pathways and food webs; and to investigate the possible effects of a radiation spill or atomic bomb on natural ecosystems.

I took a course in "radioecology," which seemed like the latest and most high-tech thing imaginable in a field whose previous high-tech devices included a map and compass. We used small amounts of radioactive chemicals and did simple experiments. We handled these materials literally at arm's length, wearing lead-lined aprons and standing on the far side of blocks of lead as we poured radioactive water from one container to another. It was still early in the development of computers and digital displays, and I was very impressed by a line of vacuum tubes, each with a number from 1 to 9 inside, that could be lit by an electric current, and sets of these, attached to a Geiger counter, that would count the number of radioactive decays. It seemed the latest in modern displays, but it actually was so simple that today's user of a Blackberry or iPhone would laugh at it.

When we were done with an afternoon's laboratory experiment, the question was what to do with the radioactive waste. It turned out the answer was simple: Following federal regulations, we simply had to wash the stuff down the sink with lots of running water. The concern was with the concentration of radiation in water, not the total amount. Thus, we could dump as much radioactive material as we wanted as long as we diluted it enough. It was a specific and legal example of the familiar cliché "Dilution is the solution to pollution," and I was quite taken aback by it. It was one small step in creating in me a certain amount of skepticism about how much faith we could have in governments to protect us from toxic substances, especially these radioactive ones.

As I previously noted, dealing with nuclear wastes is a major unsolved problem. Nations and nuclear power corporations would like you to think otherwise, as illustrated by the following quotation from a report by the World Nuclear Association (WNA) that summarizes the situation this way:[15]

- Nuclear power is the only energy-producing technology which takes full responsibility for all its wastes and fully costs this into the product.

- The amount of radioactive wastes is very small relative to wastes produced by fossil fuel electricity generation.

- Used nuclear fuel may be treated as a resource or simply as a waste. The radioactivity of all nuclear wastes diminishes with time.

- Safe methods for the final disposal of high-level waste are technically proven; the international consensus is that this should be deep geological disposal.

Dismantling nuclear power plants is part of the radioactive-waste problem

The life of a nuclear plant is just 30–40 years, and it costs more to dismantle it than to build it. Estimates for decommissioning and dismantling a large nuclear power plant run from $200 million to $500 million. Unlike power plants fueled by coal, oil, or gas, nuclear power plants have a finite lifetime because the inner workings become so radioactive that it is not possible to go in there and fix or replace things like pumps and valves, and the radiation damages the machinery so that it becomes unrepairable. In theory, a fossil fuel power plant could be run for a very long time by replacing individual mechanical parts and units as they wore out. This kind of plant is something like the ax that the old-time New Hampshire farmer had. His friend said, "Josh, that's an awfully good ax. Where'd you get it?" And Josh replied, "I've had that ax for twenty years, and all it's needed is two new heads and one new handle."

You just can't have a nuclear power plant like Josh's ax.

Is it true that the problem of nuclear waste has been "technically" solved, so we don't have to worry about it? Here are some facts. There are 441 nuclear power plant reactors in the world. A recent conference about them held in South Africa reported 220,000 tons of spent fuel—nuclear waste—worldwide since nuclear power production began in the 1950s.[16] The International Atomic Energy Agency puts it at about 300,000 tons.[17] A 2006 international conference on nuclear waste, held by the Organization for Economic Co-operation and Development's Nuclear Energy Agency, put the figure much higher, at more than 2.2 million tons.[18] This last number works out to three-quarters of a pound of radioactive waste for every man, woman, and child in the world, whether or not they had access to electricity generated by nuclear power.

That the numbers differ greatly depending on the source should be a serious concern for us citizens. Even after my years of trying, often unsuccessfully, to get good numbers about anything ecological and environmental, I was shocked that international organizations differed so widely from each other in their estimates of the amount of radioactive waste hanging around. If governments and international organizations that deal with nuclear waste don't know within a factor of 10 how much they're dealing with, how can we feel confident that they'll do a decent job of keeping us and the environment safe from it?

If you want to do the numbers on your own, here's a starting point. According to the World Nuclear Association, a nuclear power plant with 1,000-megawatt capacity generates about 30 tons of the hot stuff as waste each year.[19] This means that the Indian Point Power Plant, which has 2,000-megawatt capacity, generates about 60 tons of radioactive waste a year.

The fact is, nobody has yet worked out a good way to deal with radioactive waste.[20] According to the summary of the international conference mentioned above, "In all countries, the spent fuel or the high-level waste from reprocessing is currently being stored, usually aboveground, awaiting the development of geological repositories." In other words, the world's radioactive waste from nuclear power plants is in temporary holding facilities awaiting agreements about where on (or in) planet Earth these might be safely stored.

How long must radioactive wastes be stored before they are considered safe? A long time—exactly how long depends on which radioactive elements make up most of the wastes, since they differ greatly in the length of time each remains dangerous. But even the World Nuclear Association, which calls itself a "global private-sector organization that seeks to promote the peaceful worldwide use of nuclear power as a sustainable energy resource for the coming centuries," states that "after being buried for about 1,000 years most of the radioactivity will have decayed."[21] And this is an optimistic scenario.

According to the Alliance for Nuclear Responsibility, whose stated mission is "to protect the public and future generations from radioactive contamination" and "to provide educational materials on safety and security issues at California's aging nuclear plants," radioactive wastes from nuclear power plants remain dangerous for much longer. For example, one component of nuclear power waste is nickel-59. It loses half of its radioactivity in 76,000 years and would be hazardous for 760,000 to more than 1.5 million years, depending on how experts define "hazardous" quantitatively. Another component, iodine-129, loses half of its radioactivity in 16 million years and would be hazardous for 160–320 million years.[22] In 2002, EPA was required to create a sign that would warn people about the dangers of radioactivity at the Yucca Mountain nuclear depository for 10,000 years.[23] We can take this as the U.S. government's estimate of how long wastes from nuclear power plants remain dangerous.

What can you do with radioactive waste? Basically, there are three things. First, you can put it in tight containers, store these above-ground, and hope nothing leaks. Second, you can bury them very, very deep in the Earth and hope that the radioactive material doesn't get into subsurface water and find its way into aquifers that are then tapped by people, or reach natural vegetation, or come to the surface in natural artesian wells, springs, and so forth. Third, you can try to turn the radioactive waste into chemicals and materials so inert that they won't erode or dissolve before the radioactivity has dissipated.

Yucca Mountain

Some 70,000 tons of highly radioactive nuclear waste are stored today in a temporary facility and eventually must be moved somewhere to a safer, more permanent facility. For many years, that was going to be the Yucca Mountain nuclear repository. The plan was to move all 70,000 tons across the country to Yucca Mountain, Nevada, by truck and train: one to six trainloads or truck convoys every day for 24 years, according to the U.S. Government Accounting Office (GAO).[24] The state of Nevada pointed out that in total there would be 35,000 to 100,000 trains or truck convoys, that these would pass through many of the major metropolitan areas of the nation,[25] and at least one-third of the trains and convoys would pass through Chicago. CBS News quoted Senator Harry Reid of Nevada as saying, "Every one of these trucks, every one of these trains, is a target of opportunity for a terrorist to do bad things.... I mean, you talk about a dirty bomb. I mean this is, this is really a filthy bomb."

There are three primary temporary storage sites, and a total of 39 temporary holding facilities, many on river flood plains. More than half of the people in the United States live within 75 miles of these temporary sites,[26] and if Yucca Mountain had been used as the permanent site, at least 85% of these trains with their nuclear wastes would have passed within a half-mile of the Las Vegas strip (Figure 5.5).

In January 2010, President Obama rejected the use of Yucca Mountain as a place to deposit nuclear wastes. At the time of this writing, Steven Chu, the Secretary of Energy, had established a blue-ribbon panel to consider alternatives to Yucca Mountain, but this panel had yet to meet.[27]

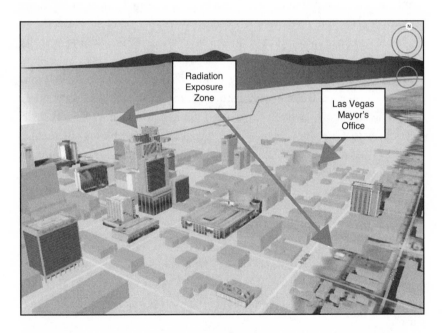

Figure 5.5 This map shows that if Yucca Mountain had been used, at least 85% of shipments of radioactive wastes from power plants would have passed within the city of Las Vegas, including near the mayor's office and within a half-mile of 49,000 hotel rooms along the strip. *(Courtesy of Fred C. Dilger PhD)*[28]

While opponents of the Yucca Mountain site will be gratified by this decision, the nuclear wastes have already been in temporary (i.e. interim) storage facilities, even though, according to the *New York Times*, these are forbidden under current law.[29] Therefore, as of this writing, there is not even a solution on the table, and the problem remains as it has been for many years. Whatever happens, a place to store nuclear waste safely is necessary for human health and the environment before the U.S. undergoes a major increase in the number of nuclear power plants.

Hasn't France solved this problem?

When I discuss my qualms about nuclear power, I am sometimes asked: If nuclear power is so bad, how come France gets almost 80% of its electricity from nuclear power, and there hasn't been a major nuclear power plant disaster there, and the French people don't raise complaints about it?

I answer this with another question: What is France doing with its radioactive waste?

France has 58 nuclear reactors. Some of the radioactive waste from these is shipped to Russia, which stores it, for a price, and is also said to process it to recover whatever usable radioactive fuel may be left in it. According to Greenpeace, France sends thousands of tons of nuclear waste to Russia, where it is processed and then stored, again according to Greenpeace, "at extremely contaminated sites in Siberia." We note that the processing, too, results in a large amount of radioactive water and materials that also then have to be dealt with.

The World Information Service on Energy (WISE) states that the French nuclear station Eurodif in southern France produces 15,000 tons of radioactive waste a year and stores 220,000 tons of waste from French nuclear facilities.[30] Notice that this figure is equal to the amount that one of the sources I found said was *the world's total amount* of radioactive waste.

France also treats some of its radioactive waste chemically, largely by using it as salts and other metal compounds in ceramics or glass, known in the trade as "vitrified waste." People familiar with the potter's wheel and kiln know that many beautiful glazes contain metallic salts, such as chromium. The radioactive stuff, much of which is metal, is ground up and mixed with clay and then high-fired, with the expectation that the resulting ceramics will last a long time. But one problem is that the production of these ceramics produces its own wastes, which are either emitted from smokestacks or dumped into water, just like the radioactive chemicals from my radioecology course. And some of this wastewater has been allowed by the French government to flow into the English Channel.

The World Nuclear Association argues that the amount of radioactive waste is not that big or bad. It points out that a typical nuclear power plant produces about 30 tons of radioactive waste annually, but when this is converted into vitrified wastes, it takes up a rather small volume, about 3 or 4 cubic yards, and that this material can then be stored in ponds at a nuclear power facility.[31] "Some 90% of the world's used fuel is stored thus and some of it has been there for decades," WNA reports. And that organization points out that the radioactive elements decay and become less radioactive over time, so that "after 40–50 years the heat and radioactivity have fallen to one-thousandth of the level at removal."

But environmental groups tell quite a different story. Greenpeace points out that France has a major nuclear-waste holding facility at La Hague in Normandy that contains 1.4 million containers of radioactive wastes. And although by French law no international dumping of radioactive wastes is supposed to occur, Greenpeace states that "an estimated 140,000 containers of nuclear waste disposed at La Hague came from foreign nuclear utilities in Europe and Japan."[32] Greenpeace claims that dairy cattle are drinking water contaminated with radioactive materials from this facility, and that even French Champagne is being contaminated by radioactive material from it.[33] Greenpeace says the French government was informed by the company managing the waste facility that a fissure had occurred in one of the storage containers due to groundwater erosion.[34]

What seems to be happening is that the containers and their surrounding facilities were constructed in the belief that groundwater would not be a powerful enough erosive force, but in fact it is.

Could nuclear power plants lead to disaster?

They could, and in a few cases a few have, primarily because anything operated by people is subject to human error. Surgeons, lawyers, presidents, generals, pilots, air-traffic controllers, engineers, scientists—you name it, we all make mistakes: A patient dies, an innocent person goes to jail, the country goes to hell, a platoon gets wiped out, planes collide, and bridges fall down. The problem with human error in managing nuclear power, and the transportation and storage of nuclear waste, is the enormity and longevity of the potential harm. The effects may be local/regional, another Chernobyl; with breeder reactors, a nuclear war; terrorism; or long-term contamination of air, land, and water due to leaks, dangerous explosions, and radioactive fallout.

Why the small investor should not invest in nuclear power

In 1957, the state of Washington started the Washington Public Power Supply System (WPPSS), which soon became known as WHOOPS, for reasons that will become obvious. It was touted as a wonderful, modern way to ensure an ample supply of electrical power for the people of the state, using the newly emerging technology of the nuclear power reactor. WPPSS, a public corporation set up by the state to build and operate nuclear plants, allowed publicly owned utilities to combine resources and build power-generation facilities.

Unbelievably, those chosen to be the directors and managers of the WPPSS system had no experience in nuclear engineering or in large projects. As a result, things went very wrong. One contractor shown to be incompetent was retained for more work anyway. Initial designs turned out to be too dangerous and unreliable. Contractors made mistakes, so some parts of reactors were rebuilt many times. Costs skyrocketed.

By the early 1980s, with not a single plant working, the cost of the entire project reached an estimated $14 billion, and the WPPSS board stopped construction. Because the nonoperating plants brought in no money, WPPSS defaulted on $2.25 billion due in bonds. And who was stuck with the bond debts? A lot of small investors who had trusted a state bond issue to be a safe way to save for retirement. In some small towns where unemployment due to the recession was already high, this amounted to more than $12,000 per customer.

There were lots of bondholders, and they sued, as did various parties involved in the design and construction, who sued each other, and the matter wound its way through the courts for 13 years. In 1988 the parties settled for $753 million. That settlement involved 30,000 bondholders, some of whom got as little as 10–40 cents on each dollar they had invested.

Two famous nuclear disasters

Three Mile Island

Although nuclear engineers and big power companies have said for years that nuclear reactors are safe, the reality is that nuclear reactors emit considerable, undesirable amounts of radiation into the environment even during "normal" operation, and when mistakes are made, they have led to some spectacular disasters.

One of the most dramatic events in the history of U.S. radiation pollution occurred on March 28, 1979, at the Three Mile Island nuclear power plant near Harrisburg, Pennsylvania. The malfunction of a valve, along with human errors, resulted in a partial core meltdown—the neat lineup of rods of uranium and graphite and the materials holding them got so hot that the whole thing melted, releasing intense radiation into the interior of the containment structure. This kind of accident was precisely what the containment structure was designed to contain, and it mostly did. Still, some radiation was released into the environment. Three days after the accident, radiation levels near the Three Mile Island nuclear power plant were six times higher than the natural background radiation. (There is always some radiation in our environment, some coming from outer space as cosmic rays, some from naturally occurring radioisotopes in the soil and bedrock.)

Since the long-term chronic effects of exposure to low levels of radiation are not well understood, the effects of Three Mile Island exposure, although apparently small, are difficult to estimate. However, the incident revealed many potential problems with the way U.S. society has dealt with nuclear power.

Since nuclear power had been considered relatively safe, the state of Pennsylvania was unprepared to deal with an accident. For example, there was no state bureau for radiation help, and the state Department of Health did not have a single book on radiation medicine. (The medical library had been dismantled two years earlier for budgetary reasons.) One of the major impacts of the incident was fear; yet there was no state office of mental health, and no staff member from the Department of Health was allowed to sit in on important discussions following the accident.

Chernobyl

The worst nuclear power plant accident occurred in 1986 at Chernobyl in Ukraine (which was then part of the Soviet Union). The World Health Organization estimates that 4,000 people have died as a direct result of this accident.[35] Experts estimate that this release will cause 39,000 cancer deaths in Europe over a period of 50 years following the accident.

Lack of preparedness to deal with a serious nuclear power plant accident was dramatically illustrated by events that began unfolding on Monday morning, April 28, 1986. Workers at a nuclear power plant in Sweden, frantically searching for the source of elevated levels of radiation near their plant, concluded that it was not their installation that was leaking radiation but that radioactivity was coming from the Soviet Union by way of prevailing winds. Confronted, the Soviets announced that an accident had occurred at a nuclear power plant at Chernobyl two days earlier, on April 26. This was the first notice to the world of the worst accident in the history of nuclear power generation.

It is speculated that the system that supplied cooling waters for the Chernobyl reactor failed as a result of human error, causing the temperature of the reactor core to rise to more than 3,000°C (about 5,400°F) and melting the uranium fuel. Explosions blew off the top of the building over the reactor, and the graphite surrounding the fuel rods used to moderate the nuclear reactions in the core ignited. Some reports suggest that the energy of the blast was 200 times greater than that released by the Hiroshima and Nagasaki atomic bombs taken together.[36] The fire produced a cloud of radioactive particles that rose high into the atmosphere. Within a short time, there were 237 confirmed cases of acute radiation sickness, and 31 people died.

In the days following the accident, nearly 3 billion people in the Northern Hemisphere received varying amounts of radiation from Chernobyl. With the exception of the 30-km (19-mi) zone surrounding Chernobyl, global human exposure was relatively small. Even in Europe, where exposure was highest, it was considerably less than the natural radiation people receive during the course of a year. In that 30-km zone around Chernobyl, however, about 115,000 people were evacuated; and as many as 24,000 people were estimated to have received a dangerous dose. We are told that this group of people is being studied carefully.

One apparent effect is that since the accident the number of childhood thyroid cancer cases per year has risen steadily in Belarus, Ukraine, and the Russian Federation (the three countries most affected by Chernobyl). In 1994 a combined rate of 132 new thyroid cancer cases were identified. Since the accident, a total of 1,036 thyroid cancer cases have been diagnosed in children under 15. These cases are believed to be linked to the released radiation from the accident, although other factors, such as environmental pollution, may also play a role. It is predicted that a few percent of the roughly 1 million children exposed to the radiation will eventually develop thyroid cancer.

Outside the 30-km zone, the increased risk of cancer is very small and not likely to be detected from an ecological evaluation. Nevertheless, according to one estimate, Chernobyl will ultimately be responsible for an additional 16,000 deaths worldwide.

Chernobyl had other environmental effects as well. Vegetation within 7 km of the power plant was either killed or severely damaged by the accident. Pine trees examined in 1990 around Chernobyl showed extensive tissue damage and still contained radioactivity. The distance between annual rings (a measure of tree growth) had decreased since 1986.

Interestingly, scientists returning to the evacuated zone in the mid-1990s found, to their surprise, thriving and expanding animal populations. In the absence of people, species such as wild boar, moose, otters, waterfowl, and rodents were enjoying a population boom. The wild boar population had increased tenfold since the evacuation of people. Still, these species may be paying a genetic price for living within the contaminated zone. A study of gene mutations in meadow voles (also called field mice) within the zone found more than 5 mutations per animal, compared with a rate of only 0.4 per animal outside the zone. It is puzzling to scientists that the high mutation rate has not crippled the animal populations, but it appears so far that the benefit of excluding humans outweighs the negative effects of radioactive contamination.

In the areas surrounding Chernobyl, radioactive materials continue to contaminate soils, vegetation, surface water, and groundwater, presenting a hazard to plants and animals. The evacuation zone may be uninhabitable for a very long time unless some way is found to remove the radioactivity. For example, 5 km from Chernobyl, the city of Prypyat, which had a population of 48,000 at the time of the accident, is today a "ghost city," abandoned, with blocks of vacant apartment buildings and rusting vehicles. Roads are cracking and trees are growing as new vegetation transforms the urban land back to green fields. Cases of thyroid cancer are still increasing, and the number of cases is many times higher for people who lived as children in Prypyat at the time of the accident.

The final story of the world's most serious nuclear accident is yet to completely unfold. Estimates of the total cost of the Chernobyl accident vary widely, but it will probably exceed $200 billion.[37]

Although the Soviets were accused of not giving attention to reactor safety and of using outdated equipment, people still wonder whether such an accident could happen again elsewhere. Because there are several hundred reactors producing power in the world today, and because given enough time human error is almost inevitable, the answer has to be yes. About ten accidents have released radioactive particles during the past 34 years. Although the probability of a serious accident is very small at a particular site, the consequences may be great. Whether this poses an unacceptable risk to society is really not so much a scientific issue as a political one involving a question of values.

Dead trees standing: a story about nuclear radiation

What is it like to be in a place that has been subjected for years to radiation of the kind a nuclear power plant or its waste would release if a spill or operating accident occurred? I can tell you what it's like because I worked as part of a team on a little-known, curious experiment conducted in the 1960s and '70s at Brookhaven National Laboratory on Long Island. There, scientists exposed a natural forest to radiation for 15 years to see what a nuclear war or the accidental or deliberate release of radioactive materials might do to one of nature's ecosystems. The experiment was done because during the Cold War the danger of a nuclear war seemed real.

So that scientists could work in the irradiated forest, the laboratory had moved the largest hunk of cesium-137 that could be safely handled by earthmoving machinery into the forest and mounted it on a vertical, movable pole that could be lowered into the ground and covered by lead shielding to protect the researchers, then raised up again automatically from a safe distance. We could work in the forest up to four hours a day because cesium's radioactive isotope-137 was relatively "clean"—only gamma rays were produced. The result appears in Figure 5.6.

FIGURE 5.6 The irradiated forest after about 10 years of exposure to intense radioactivity. (Top) From the air—the dead trees standing are visible and the bare ground beneath them. (Bottom) And near ground zero. *(Top photo courtesy of Brookhaven National Laboratory; bottom photo by Daniel B. Botkin)*

Eight years after the radiation began, I entered the forest to do my day's fieldwork, studying how the radiation was affecting the growth of trees. When all the trees are killed in an ordinary forest—say, by a disease, an insect outbreak, or a prolonged period of drought or of freezing temperatures—most of the trees soon fall over and decay. Mounds of rich organic humus, the product of that decay, are left, and soon tree seedlings and saplings sprout along with flowers and grasses adapted to open areas. Life begins anew among the dead.

But that wasn't the way Brookhaven National Laboratory's irradiated forest looked that day. The strange thing was that although the trees had been dead for years, the forest looked as if it had burned just the day before. The trees were still standing, leafless, gray and brown, because the bacteria and fungi that decompose wood were killed, too, as were the insects and worms that help with decomposition. And nothing grew anew except in small triangles of "shade" where the dead trees standing protected small patches of ground from the radioactive cesium. Behind these trees, small but hardy sedges were about a foot or two high.

A journey to the center of the forest—that is, to the source of radiation—after almost a decade of exposure was surreal. The forest was enclosed by two chain-link fences with locked gates. Just inside the fences, the woods were typical of those found on Long Island: a dense clutch of small pitch pines, scarlet and white oaks, and small shrubs, mostly blueberry and huckleberry. Many plants were quite fragrant. The sounds of crickets and cicadas filled the air. Ovenbirds called.

My walk toward the source thus began as a pleasant stroll through the woods. As I moved closer to the center, more and more pine trees had dead branches and needles. Farther on, all the pines were dead, but many were still standing. Some of the fallen pine trunks were beginning to rot—the bacteria and fungi of decay at the distance from the cesium had survived the radiation.

Up ahead, the white oaks looked sick. A bit farther on, the white oaks, too, were all dead and standing. The scarlet oaks proved to be the hardiest of the trees. As I neared the source, I saw some survivors.

It was like walking up a mountain. The higher you climb, the smaller and fewer the trees. Eventually, the trees drop out completely and you reach a zone of low shrubs, then a tundra zone of smaller ground plants, and finally, if the mountain is high enough, no visible life at all.

So it was in the irradiated forest. Blueberries and huckleberries survived the trees, growing among grasses and sedges. Closer to the source, only a patchy cover of sedges. Then you came upon perfect triangles of sedges, green, grasslike, flowering plants growing behind the trunks of standing dead trees. Just as they do with sunlight, the trunks shaded the sedges from the radiation. It was an eerie demonstration of how light rays travel.

Near ground zero, all plants were dead, but they had not decayed. The radiation had killed off the armies of decay: fungus, bacteria, earthworms, and so forth. I hunted around for any signs of life. Within about six feet of the source, I found, on the back of a sign warning of the radiation danger, a small green patch of the algae *Protococcus*, which grows on damp soils. The sandy soil encircling the source was tinted gray, the color of the dead leaves and twigs that had not decayed.

From the air, the forest was an eerily beautiful sight of death radiating outward. You could see the tower containing the radiation, surrounded by a lifeless gray-tan zone. Then came a circular ring of sedges, one of shrubs, another of oaks without pines, and then the healthy forest. Rather than the intricate mosaic of life-forms that characterizes normal forests, the pattern at Brookhaven was a series of concentric circles signifying the stages of death by radiation.

The radioactive waste generated at nuclear power plants and the problems associated with the transportation and storage of nuclear wastes can create equally mournful landscapes. For this reason, the irradiated forest at Brookhaven National Laboratory should make us pause and think carefully before we move in the direction of greater emphasis on nuclear power rather than on energy sources that are more environmentally benign.

For those who are interested: more background

What is nuclear energy?

For most of us, nuclear energy is exotic and strange, so it may be helpful to go over some of the basics here. Nuclear energy is the energy of the atomic nucleus, released by splitting atoms, a tricky business done inside what are called *nuclear reactors*. In the United States, almost all these reactors use a form of uranium oxide as fuel.

Three types, or isotopes, of uranium occur in nature. Unfortunately, the one that is useful in conventional nuclear reactors—uranium-235—is relatively rare, making up about 0.7% of the uranium found on Earth.

Most uranium—99.3% of all natural uranium—is uranium-238. The third type, also not used in conventional reactors, is uranium-234, which makes up about 0.005%. The first step in making useful uranium fuel is to concentrate uranium-235 from 0.7% to about 3%.

Uranium atoms split naturally, releasing energy, nuclear fragments, and neutrons. The neutrons go out and split other uranium atoms. Here's one of the tricky bits: If too much U-235 is brought together, lots of neutrons are produced, and there is a chain reaction that gets away. This is how an atomic bomb works, and in a reactor a runaway reaction can produce enough heat to melt the machinery and the building and emit dangerous radioactive material into the atmosphere and water. If the U-235 is not concentrated enough, not much happens, so the goal is to get just enough U-235 splitting and producing neutrons. This is done by finding a way to control the reactions: If things get going too fast, to take away the excess neutrons; if too slow, to let more neutrons fly around.

Neutrons are controlled, or "moderated," both to slow them down so they are more likely to split atoms and to control the rate of reactions. Most nuclear power plants use a combination of graphite rods and huge bathtubs of water to control and contain the reactions. Water is a good absorber of the neutrons. An eerie thing to do is to visit a bathtub reactor where you can stand on an iron grating above a very deep pool of water, as I have done. You look down and see an intense blue glow where all that atomic reaction is going on. If you were down there where the blue light is, you'd be dead. *It's sort of like looking into the devil's mouth,* I thought, standing there.

The energy from splitting uranium atoms is used to heat water and make steam, which then runs a steam turbine that generates electricity. Coal- and gas-fired electrical generators do the same thing, just using a different source of heat to boil water.

There are three kinds of nuclear reactors—conventional, breeder, and fusion—and each has its own drawbacks:

- **Conventional:** The fuel is limited; as I wrote earlier, there is only a 40-year supply or less for the world. Therefore, it is not a long-term solution.
- **Breeder:** Bombs can be made from the fuels.
- **Fusion:** This actually doesn't exist yet, and may never, so it cannot now be considered a potential solution.

The reactor is a complicated machine, full of pumps, pipes, very corrosive materials, lots of wires, and a lot of mechanisms to shut things down if something goes wrong. In the reactor core, fuel pins, consisting of enriched uranium pellets in hollow tubes (3–4m long and less than 1cm, or 0.4 in., in diameter), are packed together (40,000 or more in a reactor) in fuel subassemblies. A minimum fuel concentration is necessary to keep the reactor critical—that is, to achieve a self-sustaining chain reaction. A stable fission chain reaction in the core is maintained by controlling the number of neutrons that cause fission. The control rods, which contain materials that capture neutrons, are used to regulate the chain reaction. As the control rods are moved out of the core, the chain reaction quickens; as they are moved into the core, the reaction slows. Full insertion of the control rods into the core stops the fission reaction.

The coolant removes heat produced by the fission. The rate of heat generation in the fuel must match the rate at which heat is carried away by the coolant. All major nuclear accidents have occurred when something went wrong with the balance, allowing heat to build up in the reactor core.

In a meltdown, the nuclear fuel becomes so hot that it creates a molten mass that breaches the walls of the reactor and contaminates the environment.

Nuclear power reactors, each of which produces about 1,000MW of electricity, require an extensive set of pumps and backup equipment to ensure that adequate cooling is available to the reactor. Smaller reactors can be designed with cooling systems that work by gravity and are thus not as vulnerable to pump failure caused by power loss. Such cooling systems are said to have *passive stability*, and the reactors are said to be *passively safe*.

The bottom line

- Conventional nuclear power plants are not a short-term solution to the energy problem, because they take too long to build. They are also not a long-term solution, because nuclear plants have short lives and their fuel is a rare mineral that will run out in decades.

- Their radioactive wastes, however, will be with us for thousands of years, and there is no satisfactory solution to dealing with these wastes.

- Although governments and international agencies say nuclear waste is being handled safely, evidence suggests the contrary. In particular, ground water is more erosive than engineers expected and has corroded some facilities, allowing radioactive water to escape into the ground and then contaminate surface water.

- Nuclear power is expensive, but many of the costs are indirect and thus not evident: for example, development of atomic reactors has been funded primarily by governments, as is dealing with the wastes and the cleanup when contamination occurs.

- No insurance company will insure a nuclear power plant. As a result, governments are responsible for any damage and law-suits—another hidden cost.

- Nuclear power plants have had only a few major accidents, but these have been costly.

- Exploration and prospecting for uranium ore will probably increase greatly, especially in remote, relatively unexplored areas, raising new problems for conservation of biological diversity, along with environmental pollution and, locally, human health.

- In sum, nuclear energy is a cure worse than the disease. And since there are alternative sources of abundant energy that don't pose the hazards of nuclear power, why take the unnecessary risks that it entails?

Section II

New energy sources

Americans built great public work projects to pull ourselves out of the Great Depression. The hydroelectric projects on the Columbia, the Colorado, and the Tennessee rivers are witness to what we can accomplish when we put our minds to it. We rose to the challenge of fascism in WWII. We belatedly granted civil rights to all our citizens in the 1960s and in the modern era we have pushed cigarette smoking to the fringes of society.

Our next great challenge will be the rapid conversion of American electricity supply from fossil fuels to renewable sources of energy, and the conversion of the bulk of personal transportation to electric vehicles. In doing so we can transform society and reindustrialize the continent's heartland.

—Paul Gipe, wind energy expert

If it makes people feel good to shove up a windmill or put a solar panel on their roof, great, do it. It'll help a little bit, but it's no answer at all to the problem.

—James Lovelock, internationally known environmental scientist and father of the Gaia hypothesis about how all life on Earth is connected

If experts like these two can disagree as much as they do, it's no wonder that you may be confused about whether alternative forms of energy can meet our needs. In this section, we explore the major sources of renewable alternative energy. By the end of this section, you can better choose between the assertions of the wind energy expert Paul Gipe and the internationally famous environmental scientist James Lovelock, and understand which are the better alternatives. Should we turn to wind, solar, ocean, and biofuels—or perhaps a combination of these and other energy sources? If so, what would be the best combination?

To make a start at considering these alternative sources, we need to look again at a chart you saw earlier, showing world energy use and how much each of these alternatives contribute today worldwide (Figure SII-1). This helps you to see how much they would have to grow to become major energy sources that could replace fossil fuels to a significant extent.

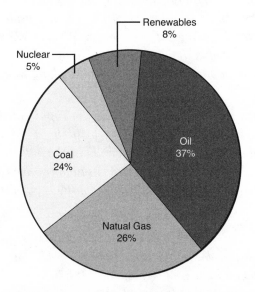

FIGURE SII.1   World energy use 2010 by fuel type

In the United States, alternative renewable energy sources, a group that for our purposes includes wind, solar, and biofuels, today provide 3.7% of the energy, but biofuels, taken together, are the largest contributor. These include wood cut and burned, waste such as cooking oil that is later burned, and crops grown to be converted directly into fuel. *All* renewable energy sources, which typically include the three just mentioned—wind, solar, and biofuels—as well as freshwater energy and geothermal energy, provide 7% of the energy in the United States. If we focus only on solar, wind, and biofuels, we find that firewood provides about half, crops grown as fuels 23.3%, and wastes 10%. Solar and wind are a small part of this group, 7% and 2%, respectively (Figure SII.2).

This suggests that all the renewables together—conventional and alternative—play rather small roles in energy supply today. When, if ever, will renewable energy provide most, or at least half, of the world's energy? And at what costs, both economic and in terms of undesirable effects on the environment and the lives of people?

We can see the beginning of an answer in the fact that use of renewable energy is increasing. Between 2005 and 2006, it increased 7%, while *total* energy use actually declined 1% (mainly due to reduced use of fossil fuels).[1] But if we continue to move away from fossil fuels, will we face a future in which our energy supply and per-person use is so limited that we must accept a decline in our overall standard of living? This section helps to answer these questions.

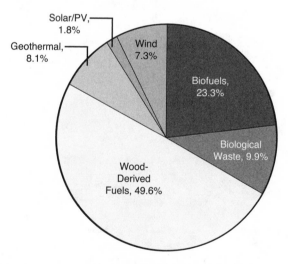

FIGURE SII.2   U.S. renewable energy 2007[2] *(Source DOE EIA)*

6

Wind power

FIGURE 6.1 Under a hazy sun, a pair of pronghorn antelope graze in a west Texas field as wind turbines generate electricity in the background. *(©iStockphoto.com/chsfoto)*[1]

Key facts

- Wind is the cheapest alternative energy source. Electricity from wind energy costs between 4.5 cents and 7.5 cents per kilowatt-hour, making it cost-competitive with fossil fuels.

- The windiest 20 states have enough wind energy to potentially provide one-third to one-half of the total U.S. energy use, and all of its electricity, now and in the next 40 years.
- Today's wind turbine is as tall as a 22-story building and generates enough electricity for 500 modern U.S. homes.
- Each of these modern wind turbines cost about $2 million to install, all costs included.
- Worldwide, wind power provides just a few percent of the electricity, but installations are increasing rapidly, and in some nations wind provides as much as 40% of the energy. Germany gets 5% of its electricity from the wind.
- The best sites for wind power in the U.S. are along the coasts, including just offshore and in the coastal mountains of the Pacific states, and also in the Midwest, especially in a north-south belt stretching from eastern Montana and Minnesota to the Texas panhandle.
- Wind energy is also a great benefit in the developing nations and rural areas, and for the poor, especially where other forms of energy are not easily available. In these situations, small windmills that generate electricity are becoming an important part of self-help to raise standards of living.
- In the U.S., almost 7,000 small windmills are sold each year for generating electricity on farms and for individual homes.

Sailboats and windmills are ancient

Opinions about wind energy have blown hot and cold, variable as the wind itself, viewed in some centuries as important and fashionable, in others not so much. A drawing of a sailboat appears on an Egyptian pottery vessel 4,000 years old. The earliest known windmills were Persian, dated as early as 900 BC.

Along with watermills, windmills were important sources of energy in the Middle Ages and the Renaissance. They continued to be important on a large scale into the early part of the scientific-industrial age in the 19th century, and locally, mainly in rural areas, they remained important for pumping water for some time into the 20th century. One history of wind power states that "between 1850 and 1970, over six million mostly small (1 horsepower or less) mechanical output wind machines were installed in the U.S. alone,"[2] pumping water for cattle and people. They are familiar today as decorative, if no longer functioning, landmarks of the rural countryside.

The first windmills to generate electricity were built at the end of the 19th century. They were overshadowed by the invention of the coal-powered steam engine and petroleum-powered gasoline and diesel engines, which could run almost anywhere, anytime, a convenience that soon came to be considered a necessity and just couldn't be beat. Few today know about the first windmill to generate electricity, which was built in 1888 in Cleveland, Ohio. Know as the Brush postmill, it made use of modern metals, which provided greater strength and allowed greater size than the earlier picturesque wind machines. Its "pinwheel" blades had a diameter of 56 feet (17 meters). It was new and impressive, but it couldn't compete with coal, oil, and steam engines, and this kind of machine did not become popular.[3]

Winds of change have freshened quickly for wind power in the 21st century, with renewed enthusiasm for it perhaps best illustrated by a completely new kind of sailing ship, the MV *Beluga SkySails*, which completed its maiden voyage of 11,952 nautical miles on March 14, 2008, sailing from Bremen, Germany, to Venezuela and back carrying heavy industrial equipment (Figure 6.2).

The *Beluga SkySails* is novel in two ways. First, it did not use a set of fixed masts with traditional sails that had to be monitored constantly and their settings changed with each variation in the wind, thereby requiring either a large crew or sophisticated and expensive equipment. Instead, this ship flew a huge kite that spread out over more than 160 square yards. Flying high in the sky and attached to the hull by stout cables, it caught more reliable winds aloft than occur at the surface and required only a flexible cable to tether the kite-sail to the ship (Figure 6.2).

Second, and almost as important, wind provided only part of the power—on the maiden voyage 20%—the rest coming from standard marine diesel engines. With the *Beluga SkySails* it isn't an either-or situation, either alternative fuel or fossil fuel, but an integration of several energy sources. Diesel provided the steady energy, while the kite-sail helped when wind was available, reducing fuel costs by $1,000 a day. With this saving and also saving the expense of a large crew, *Beluga SkySails* was an economic success. At the end of the voyage, the ship's captain, Lutz Heldt, said, "We can once again actually 'sail' with cargo ships, thus opening a new chapter in the history of commercial shipping."

FIGURE 6.2 Wind power is becoming so popular that it is even making a comeback as a way to propel ships. Here a new kind of sail, actually a cable-tethered kite, helps pull a new ship through the ocean. *(Copyright SkySails)*[4]

It's worth repeating two key insights that come from this design, which are important as we work out how we are going to develop energy systems (arrangements that use a variety of sources of energy together) and turn more and more to alternative energy. The first insight is simply to integrate different forms of energy, making use of the strengths of each. In this ship's case, the combination is the reliability of internal combustion engines burning fossil fuel (or possibly in the future a bio-fuel), along with free wind energy. The second insight is the kite design itself, which instead of working against the natural forces, like the rigid masts of a sailing ship, is flexible, giving, shifting, and moving with the wind even as it helps pull the ship through the water. Therefore, it is much less likely to be damaged by storms.

Stephan Wrage, the SkySails company's managing director, said, "In the future, depending on the route and weather conditions, we'll be able to post fuel savings of between 10% and 35% using wind power." The ship is the product of private industry, co-funded with 1.2 million euros from the European Union's "LIFE" program.[5]

Can wind energy be a major player in the United States or on the world stage?

SkySails is intriguing, but does it really signal that wind power can be an important part of our energy supply? Or is it another appealing curiosity?

With worldwide concern over global warming, some people feel that the only question of interest is the global one: What kind of energy will make a dent in worldwide use of fossil fuels and the release of greenhouse gases? But many people in the world live where energy is scarce, expensive, and a fundamental cause of poverty. For them, even if an alternative source of energy can't solve the entire world's problem, it might be valuable locally. I consider both possibilities.

The largest wind energy installation in the United States east of the Mississippi is the Mountaineer Wind Energy Center in Tucker and Preston counties, West Virginia, owned and operated by Florida Power & Light and producing electricity since 2002. It has 44 big machines, which, by the way, are no longer called by their traditional and humbler name, *windmills*. These are *wind turbines*, each 228 feet high, the height of a 22-story building. Together, these turbines have the capacity of 66 million watts and in a typical windy year can produce 170 million kilowatt-hours, which is enough electricity for 22,000 homes. Each turbine is installed on 100 acres, and the entire facility takes up 4,400 acres, or just under 7 square miles.[6, 7] There are larger wind farms in the Midwest and Far West, such as one of the first, the Tehachapi Wind Farm in California (Figure 6.3).

FIGURE 6.3 Here are some of the 5,000 wind turbines on the Tehachapi Wind Farm in California, east of the San Francisco Bay, and one of the largest collections of wind generators in the world. Several private companies share ownership. In total, these turbines produce enough electricity for 350,000 people. *(DOE Photograph http://www.doedigitalarchive.doe.gov/SearchImage. cfm?page=search)*[8]

Wind energy potential in the United States

According to the American Wind Energy Association, the windiest 20 states have wind energy to potentially provide one-third to one-half of the U. S. total energy use, and two and one-half times as much energy as all of present electricity generated. Could enough wind turbines be manufactured in the United States? One analysis concludes that

> for the wind to generate ~1,000 TWh/yr, we would need to install ~500,000 MW of wind generating capacity across the breadth of the country....Americans drive ~5,000 billion km/yr. To power this fleet with electric vehicles would require a huge new supply of clean electricity. Current electric vehicles can travel ~0.25 km per kilowatt-hour of electricity supplied. Thus, converting passenger vehicles to electricity will

require the generation of ~1,000 TWh/yr. Using the same assumptions as before, this would demand the installation of ~500,000 MW of new wind generating capacity....There's more than ample land area in the US for such a large number of wind turbines. Even with a very open spacing, for example 8 rotor diameters by 10 rotor diameters apart, ~1 million MW would require little more than 3% of the land area of the lower 48 states. And of this land, the wind turbines would only use about 5% for roads and ancillary facilities.

Moreover, the US has the manufacturing capacity to build such a large number of machines within less than two decades.

Every year American manufacturer's of heavy trucks churn out ~300,000 vehicles. Each heavy truck is the equivalent of a 500,000 watt wind turbine. Thus, heavy truck manufacturers alone build the equivalent of ~150,000 MW/yr.

If two-thirds of truck production were diverted to manufacturing wind turbines, the industry could build ~100,000MW/yr. Thus, it is theoretically possible that the American heavy truck industry could provide 1,000,000 MW in about one decade.

Clearly one million MW of wind capacity in the United States alone is an ambitious target, but it's a target worthy of a great nation.[9]

The largest wind energy facility in the world is the Horse Hollow Wind Energy Center near Abilene, Texas, owned and operated by Florida Power & Light. It has 421 wind turbines with a total generating capacity of 735 megawatts, enough to meet the electricity needs of approximately 220,000 homes. Figuring an average of about three people per household, this would be enough electricity for all domestic use in Austin, Texas, a city of about 650,000.[10] The Horse Hollow wind turbines are spread widely across approximately 47,000 acres, so the land is used for ranching and energy production.[11, 12]

FPL states that wind energy installation costs $1.5–2.0 million per megawatt, in the same cost range traditionally estimated for a coal-fired power plant, but about twice the $800,000 cost of electrical generators powered by natural gas.[13] (Note, however, that in Chapter 3, "Coal," I point out that the cost of building coal-fired power plants is increasing rapidly, with estimates as high as $3.50 per installed watt.) Thus wind energy is cheap, working out to about $1.50–2.00 to install each watt of capacity, or about $150–200 to install enough generating capacity to run a 100-watt lightbulb indefinitely. And of course once installed, wind turbines require no fuel purchases, so FPL can produce electricity from wind energy for between 4.5¢ and 7.5¢ per kilowatt-hour.[14,15,16]

U.S. fossil-fuel electrical power plants have a total capacity of 754,989 megawatts. With installation costs of $2 million a megawatt, it would cost about $1.5 trillion to replace all fossil-fuel electrical production with wind energy. But consider this: According to Nobel Laureate Economist Joseph E. Stiglitz, the true cost of the Iraq war will be $1–2 trillion, based on the federal budget office's acknowledgment that $500 billion has already been spent and taking into account the lifetime health care of severely wounded soldiers, among other factors.[17] Thus, for the cost of the Iraq war, the United States could have installed enough wind turbines to produce all the nation's electricity. I discuss this in greater detail in Chapter 13.

This, of course, is subject to the major technological constraint on wind energy: that the wind bloweth where it listeth—it isn't windy all the time. One of the crucial issues is therefore storage and transportation of energy, a problem that also confronts nuclear power plants (which do best if they operate at maximum output all the time), water power, and solar energy. There are solutions, which we explore in detail in Chapter 13. The *Beluga Sky Sails* points to one solution: Use fossil fuels and nuclear power to even out the energy supply, moving to wind and other alternative sources for the majority of our energy. It is worth pointing out here that the simplest solution is to improve the electrical grid and distribute wind turbines widely, since when it's calm in one location, it's likely to be windy in another.

The answer to the question of whether wind energy can play a major role on the world stage is yes. Since we know it can provide all the electrical energy that the world's greatest energy user needs, obviously it can be a major player. But here's some additional information about the status of wind energy worldwide.

Global wind energy capacity

Globally, the wind-energy picture is similar to that in the United States, with total wind-generated electrical capacity growing rapidly (Figures 6.4 and 6.5). Wind energy is used in more than 70 nations, with the greatest use in the United States, Spain, and China.[18] It provides just 5% of Germany's electricity, but in some nations it provides 40% of the energy.[19] By 2007, the world's wind-energy generation had reached 93.8 million kilowatts, producing 200 billion kilowatt-hour per year. However, this is only 1.3% of global electricity consumption.

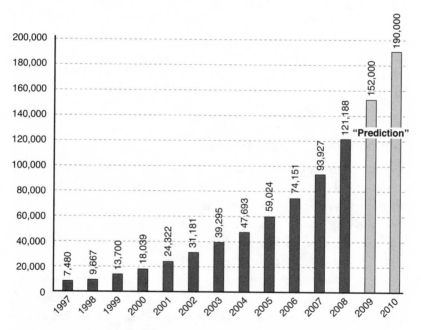

FIGURE 6.4 Total worldwide installed wind energy capacity.[20] *(World Wind Energy Association (WWEA), www.wwindea.org)*

Top 10 Total Installed Capacity			Top 10 New Capacity		
Country	MW	Percent	Country	MW	Percent
Germany	22,247	23.6	United States	5,244	26.1
United States	16,818	17.9	Spain	3,522	17.5
Spain	15,145	16.1	China	3,449	17.2
India	8,000	8.5	India	1,730	8.4
China	6,050	6.4	Germany	1,667	8.3
Denmark	3,125	3.3	France	848	4.4
Italy	2,726	2.9	Italy	603	3.0
France	2,454	2.6	Portugal	434	2.2
UK	2,389	2.5	UK	427	2.1
Portugal	2,150	2.3	Canada	386	1.9
Rest of World	13,019	13.8	Rest of World	1,726	8.6
Total top 10	81,104	16.2	Total top 10	19,350	91.4
Total	**94,723**	**100.0**	**Total**	**20,076**	**100.0**

FIGURE 6.5 Top ten nations in wind power. *(Source: World Wind Energy Association website, http://www.wwindea.org/home/index.php?option=com_ content&task=view&id=198&Itemid=43&limit=1&limitstart=3)*

How far along is wind power?

In recent years an important step was taken in developing wind energy by assessing the wind-energy potential of each area of the United States (see the later Figure 6.6). In the United States in the late 20th century, interest in wind power, as well as the potential for it, was concentrated in California and dismissed broadly elsewhere as another one of those oddball California fads, perhaps talked about on Haight Street in San Francisco or at weekend concerts in Golden Gate Park. Wind energy was believed to be great enough for commercial electrical production only along coasts or in mountain passes that funneled between hot deserts and the cool Pacific in California, Oregon, and Washington. Thus, two of the places where modern wind turbines were first installed, in the 1970s and 1980s, were Altamont Pass, east of San Francisco near Livermore, California, and Tehachapi Pass, east of Los Angeles (Figure 6.3).

Two start-up companies, Zond and U.S. Windpower (later Kenetech), dominated 1980s wind energy development in the United States, competing with Danish and Japanese machines. Unfortunately, the U.S. approach suffered from three problems. The first was that few people took wind power seriously. Second, the machines were underdesigned, not strong enough for powerful winds and storms. And third, wind power engineers believed that this kind of energy would be limited to coastlines and areas with unusual topographic features that focused the wind, like the canyons mentioned above between the hot deserts of eastern California and the cool Pacific shore.

Although wind machines of that time looked sleekly modern, the public didn't realize there was anything else radically different between them and the farm windmills of the Great Depression that pumped water for poor rural people. However, there was a hint of something different at Altamont Pass, where hundreds of large windmills were controlled by a roomful of first-generation Apple computers. Indeed, the potential for wind energy in the modern technological age stemmed as much from computers and solid-state electronic devices—along with some new materials, including fiberglass and carbon fibers—as it did from the conviction of 1980s environmentalists that we had to move away from fossil fuels.

A serious problem with modern wind machines is that they cannot completely shut themselves down automatically in high winds. They have a complex set of brakes, transmissions, and computers that slow and try to stop the blades when they spin too fast, but it is always possible that the machines will "run away," meaning that the winds will become so powerful so quickly that the braking mechanisms don't have time to work. When that happens, the turbines spin faster and faster, destroying the braking mechanisms and eventually blowing themselves and the entire structure apart and scattering their huge blades over the countryside. At U.S. WindPower, when a machine ran away, everybody just cleared out until the machine destroyed itself—it was too dangerous to stick around.

The problem was well known, but U.S. WindPower nevertheless tried to save money and underbuilt its machines. As a result, too many blew apart at the installations, and by the 1990s the company went out of business. Zond, mostly an installer, not a designer, of wind machines, held on for a while but was bought out in 1997 by Enron, and its demise became part of Enron's tragedy.[21]

After the demise of Zond and U.S. WindPower, no major American company focused on windmills. Boeing and General Electric had a few government contracts and built a few gigantic, unwieldy machines. One on the east coast of Oahu in Hawaii was so huge and heavy that its blades hardly seemed to move—you had to stare at it for a while to be sure they were turning. Understandably, these never led beyond government research contracts.

Although wind energy went through a dry spell in the United States, Japanese and Danish companies continued to build and improve the new computer-based, light-material machines, making inroads in the U.S. Once the center of science, engineering, and technological innovation— Fulton's steamboat, Edison's lightbulb, Bell's telephone, the Wright brothers' airplane, the first atomic reactor, the first windmill to generate electricity, and one of the first large hydroelectric facilities—the U.S. disappeared as a player in one of the great technological opportunities at the end of the 20th century. A growing market, a likely source of profit, was lost to us, in good part because Americans simply did not take wind power seriously. The big oil corporations didn't help—we can only surmise, not prove, that this was because it was not in their interest for Americans to move away from fossil fuels.

Then an amazing thing happened. Wind energy turned out to be cheap, its electricity economically competitive even with coal-fired power plants. And engineers found, to their surprise, that California did not have a corner on windiness in North America, nor did coastal areas. If they'd been talking to ranchers and farmers in the Dakotas and Texas, they would have realized sooner what all the ranchers and farmers knew—that the western Great Plains are a windy place. Farmers could have told them the old story that it's so windy in North Dakota that when the wind stops all the chickens fall over. Specifically, new studies showed that wind energy is also high in the western states just east of the Rocky Mountains—eastern Montana, Nebraska, Colorado, and New Mexico— and in the Midwest, especially Minnesota, North and South Dakota, Missouri, Oklahoma, and the Texas panhandle (Figure 6.6 and Table 6.1).

Between 1999 and 2008, wind power generation capacity in the United States increased sixfold to more than 16,000 million watts.[22] More than 30 states have wind turbine installations, and some of the largest of these are in Texas. The two largest are the Horse Hollow Wind Energy Center, in the countryside near Abilene, Texas, with a capacity of 735.5 megawatts, and the recently completed Roscoe Wind Farm near Roscoe, Texas, with a capacity of 781.5 megawatts. The Sweetwater Wind Farm in the southern end of the panhandle has about two-thirds the capacity of Horse Hollow. Other large ones in Texas are the King Mountain, and Desert Sky.

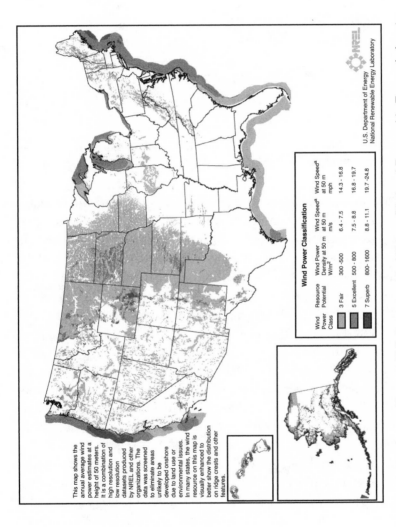

This map shows the annual average wind power estimates at a height of 50 meters. It is a combination of high resolution and low resolution datasets produced by NREL and other organizations. The data was screened to eliminate areas unlikely to be developed onshore due to land use or environmental issues. In many states, the wind resource on this map is visually enhanced to better show the distribution on ridge crests and other features.

Wind Power Classification

Wind Power Class	Resource Potential	Wind Power Density at 50 m W/m²	Wind Speed[a] at 50 m m/s	Wind Speed[a] at 50 m mph
3	Fair	300 -500	6.4 - 7.5	14.3 - 16.8
5	Excellent	500 -800	7.5 - 8.8	16.8 - 19.7
7	Superb	800 -1600	8.8 - 11.1	19.7 -24.8

U.S. Department of Energy
National Renewable Energy Laboratory

FIGURE 6.6 U.S. wind resources map (*Source: U.S. Department of Energy/National Renewable Energy Laboratory, www.eere.energy.gov/windandhydro/windpoweringamerica/*)

TABLE 6.1   The Top Ten Wind Power States (Megawatts Capacity)

	State	Existing	Under Construction
1	Texas	4,356.35	1,238.28
2	California	2,438.83	165
3	Minnesota	1,299.75	46.4
4	Iowa	1,273.08	116.7
5	Washington	1,163.18	126.2
6	Colorado	1,066.75	0
7	Oregon	885.39	15
8	Illinois	699.36	108.3
9	Oklahoma	689	0
10	New Mexico	495.98	0

(Source: www.awea.org/projects/)

California also continues to have some of the biggest wind installations, including the Tehachapi Pass Wind Farm (690 MW); San Gorgonio Pass Wind Farm, almost as big, in Palm Springs (which claims to be the only major wind farm that offers tourists a guided tour); and the famous Altamont Pass Wind Farm near Livermore.

New York State has one very large installation, the Maple Ridge Wind Farm (198 MW) in Lowville, west of the Adirondack State Park, east of Watertown and Lake Ontario.[23]

Many European nations are committed to wind energy, including Britain, Denmark, Germany, Italy, Japan, the Netherlands, Norway, Spain, and Sweden. Denmark, where many of the world's wind turbines are manufactured, has some of the world's largest installations, including Nysted Wind Farm, a joint Danish–Swedish project, the largest offshore wind farm in the world, and Horns Rev (160 MW).[24] Also impressive is Spain's use of wind energy. A major milestone was reached for world wind energy on April 19, 2008, when wind produced more than one-third of Spain's total electricity production; nuclear power was second.[25] Germany generates 5% of its electricity with wind turbines.[26] Elsewhere, Australia has a large installation, the Waubra Wind Farm,

In sum, the technology is ready to harness the Earth's cheapest alternative energy source, and installations are increasing, but there is a long way to go before wind becomes one of America's and the world's major sources of electricity. The potential is there, waiting for investments.

Wind power for rural areas, for the poor, for single-family homes, and for less-developed nations?

With all the concern about global climate change and the need to find global solutions, the usefulness of alternative energy sources for the rural and the poor is often forgotten. But people in many parts of the world lack the easy access to energy that those who live in industrial nations take for granted—as someone has said, "The people in that country were so poor, they didn't even have cell phones!" (And to think, on top of that, they didn't have easy access to electricity either.) As I discuss in the Chapter 9, "Biofuels," many African nations lack such access, and obtaining energy for cooking and heating and for minimal healthiness is difficult and time-consuming. Wind energy is rescuing some of these individuals.

In rural Malawi, 20-year-old William Kamkwamba used diagrams he found in an old book to build three windmills, using plastic pipes that he flattened for blades, a basic structure of wood from local trees, and moving parts from bicycles (Figure 6.7).[27] One of his windmills was 39 feet high, as high as a four-story building. It powered ten small lightbulbs, a TV set, and radio, all of which made a major change in his life and his family's. Not only could his sisters study for school late into the night, but his windmills became a local attraction and have taken him from a hobby to his primary work. He went on to build a windmill large enough to pump water for his village of 60 people and another to provide electricity for the local school. At his home he then added a manufactured windmill and photovoltaic solar panels, and he is putting another manufactured windmill in Lilongwe, Malawi's capital.

Kamkwamba describes how he built the first windmill. "My problem was that I didn't have much money to buy parts to construct the windmill. Over time, I found materials that had been discarded by other farmers or by the nearby tobacco plantations, and I bought a few parts with money I scraped together: 500 Kwacha (Malawian currency) or $2.75 (US $1=145 Kwacha) for two bearings; 500 Kwacha for a bicycle dynamo (the kind that powers a bike's light when you ride the bike); 400 Kwacha for a fan belt; 800 Kwacha for a bicycle frame." In sum, he spent about $15.

FIGURE 6.7   William Kamkwamba and a windmill he built in Malawi.
(© 2009 Tom Reilly, www.movingwindmills.org)[28]

This is the kind of self-help project that economists concerned about development want to see. Thinking back to Chapter 4 on water power and the 20th-century emphasis on large, expensive, and centralized hydroelectric dams, you might ask why the World Bank, whose stated purpose is to alleviate poverty and promote economic development, isn't spending its money on these relatively inexpensive local projects initiated by individuals, rather than on large projects such as the Nam Theun 2 hydroelectric dam in Laos. That dam, you will remember from Chapter 4, is displacing local residents rather than helping them and is primarily going to provide electricity not for Laotians but for export.

I asked that very question back in the 1980s when I was a consultant to the World Bank concerning biological diversity. I was told by a Bank official that it had a problem dealing with small projects since its staff couldn't get job promotions if they were bogged down handling a lot of small stuff that didn't amount to much on their resumés. What did they consider a small project? Anything under $100 million.

Mr. Kamkwamba has been getting some help from a comparatively small entrepreneurial group that calls itself Technology Entertainment Design. His story, although unusual and impressive, is not unique. Wind and solar energy offer remarkable new potential for self-help in undeveloped nations.

And don't count the United States out of the market for small wind turbines—less than 100 kilowatts. In 2005, small windmills provided a total of 30 megawatts generating capacity (only about half a percent of the total generating capacity of fossil fuels). In 2007, 6,807 small windmills were sold in the United States, adding a total generating capacity of 17 MW. Buyers paid a total of $56,082,850 for these, an average of $8,239 each, or $3,197 per installed kilowatt.

This market is growing 14–25% a year. The payback period—how long it takes for the value of the energy generated to equal the amount invested in purchase and installation—varies from 6 to 30 years for small wind turbines, depending on how good the location is for wind energy, the costs of purchasing energy commercially, permitting costs, and the energy efficiency of the turbine. Naturally, the cost per kilowatt-hour is greater for small wind turbines than for larger machines on large wind farms since the latter benefit from economies of scale. Current costs of generated electricity for these small wind turbines are about 10¢ to 11¢ per kilowatt-hour—about twice the cost of electricity produced on one of the large wind farms.

According to the American Wind Energy Association, the major markets for small wind turbines in the United States are in California, New York, Vermont, Massachusetts, Pennsylvania, Ohio, Texas, Maine, and Arizona.

> [These states have favorable] policies, regulations, and incentives [that] are conducive to making small-wind installations practical, cost-effective, and simply feasible. Costs of traditional electricity and wind resource quality are also important factors…. However, without conscientious zoning, permitting, and grid-interconnection regulations in place, rebate and incentive programs have no platform upon which to grow the

market and streamline production. Small Wind is poised to play an important role globally to supply clean, affordable, and local power to end-users, and Small Wind's long history shows that the industry is capable of overcoming obstacles it faces. Geopolitical, climatic, and economic forces ... will continue to drive demand, but the industry's maturation and accelerated growth are largely dependent on U.S. policy at the national, state, and local levels. As the U.S. lags behind world competitors in other renewable energy technologies, Small Wind remains the only major U.S.-dominated clean energy technology. With the right policies in place, Small Wind could lead the U.S. into a stronger position in the global energy paradigm.[29]

And finally, there are *microturbines*, which generate less than 1 kilowatt but do so where it is otherwise difficult or impossible to generate electricity, and these are used to charge batteries on boats and to pump water on farms and for individual houses.[30]

Downsides: wind power and the environment

No energy source is without any drawbacks whatsoever. Wind power has only a couple: Some people don't like the way the machines look—especially the large wind turbines, and most especially when the turbines are smack in the middle of a scenic vista—and sometimes birds fly into them and are killed. These two negatives can cause a lot of controversy.

In August 2009, North Carolina's state senate approved a bill that bans windmills larger than 100 feet from windy ridges in the state's scenic western mountains. (At the time of this writing, the bill has still to go to the state House of Representatives.) According to the *New York Times*, "Advocates of wind energy say that if the ban becomes law, it could put as much as two thirds of North Carolina's onshore wind resource offlimits."[31]

Not in my backyard

The Cape Wind project, proposed by Cape Wind Associates, involves placing 130 horizontal-axis wind turbines, each higher than the Statue of Liberty, on floating offshore platforms distributed over 24 square miles (therefore not very close together) in Nantucket Sound. The turbines are expected to generate enough electricity for 420,000 homes, but although originally proposed in 2001, work on the $1-billion project has yet to begin. The federal government has received more than 40,000 comments about it, the greatest number ever about an alternative-energy project,[32] and in response has held 50 public hearings. Fishermen have expressed concern about possible effects on fish habitats, including spawning grounds. Shippers have expressed concern about possible dangers for commercial shipping. And some residents of Cape Cod, Martha's Vineyard, and Nantucket have opposed the project on aesthetic grounds—that the wind turbines will spoil the scenery.[33] The late Senator Edward M. Kennedy, Democrat of Massachusetts, was one of the leaders in the fight against it, as is his nephew, Robert Kennedy Jr.[34]

It is worth noting that in the meantime, a Dutch company, Blue H, USA LLC, announced plans in March 2008 to put 120 turbines on floats 23 miles south of Martha's Vineyard and 45 miles west of New Bedford, farther away from the view of residents.[35]

Further complicating matters, the National Park Service says that Nantucket Sound is eligible for listing on the National Register of Historic Places. And two tribes—the Mashpee Wampanoag of Cape Cod and the Aquinnah Wampanoag of Martha's Vineyard—make two claims against the development: that it might be built over ancestral burial grounds and might interfere with a ritual that requires an unobstructed view of the sunrise.[36]

The aesthetic concerns about wind turbines are legitimate and pose a dilemma for environmentalists, who need to choose between the national advantage of wind power and the local effects of wind turbines on landscape beauty. Proponents of other sources of energy, such as nuclear power advocate Senator Lamar Alexander, use this aesthetic argument as a reason to oppose wind energy and support nuclear energy.[37]

Birds and wind turbines

Wind turbines can usually be placed where they won't interfere with spectacular scenery, but no matter where you put them, you can't stop birds from flying into them. You can, however, put the wind turbines in places that birds do not use very much, or you can make a mistake and put the turbines right in the middle of a bird species' nesting grounds or flyway.

The 5,000 wind turbines in Altamont Pass, California, famous among wind-power enthusiasts as one of the places that started it all in recent times, killed about 1,000 birds in one year. Half of these were birds of prey, including 24 golden eagles, a species protected under the Endangered Species Act.[38] This works out to about 1 bird killed a year for every 10 of the large windmills running. The California Center for Biological Diversity filed a lawsuit in 2004 against the operators of the Altamont Pass wind turbines, who in response agreed to remove 100 turbines that appeared to be in problem locations and to shut down half the turbines in winter months, when birds are most likely to fly into them. In addition, five Audubon chapters, including Golden Gate Audubon, filed another lawsuit against the county, claiming that there had been inadequate environmental review.

Despite the problems at Altamont Pass, the National Audubon Society has endorsed wind power, arguing that on balance its environmental benefits outweigh its disadvantages, and that it has much less effect on bird mortality in other locations, partly because of the terrain and topography of Altamont Pass and because small mammals that are the primary prey of large raptors are especially abundant there. According to the National Audubon Society, "Farms and ranches of the West and Midwest are now favored homes for wind turbines, and so far they seem to be relatively safe for both raptors and songbirds. 'The bird mortality we're seeing is lower than what's been seen at Altamont,'" says Tim Cullinan, director of science and bird conservation for Audubon Washington and a wildlife biologist."[39]

He went on to say, "We can't lose sight of the larger benefits of wind…. The direct environmental impacts of wind get a lot of attention, because there are dead bodies on the ground. But nobody ever finds the bodies of the birds killed by global warming, or by oil drilling on the North Slope of Alaska. They're out there, but we don't see them."[40]

The most notorious problem with wind turbines killing birds is at the Norwegian wind farm at Smola, a set of islands six miles off the northwest coast of Norway. In 1989 Birdlife International listed these islands as an important area for conservation of birds because they had one of the world's highest densities of white-tailed eagles, close relatives of the North American bald eagle. Ignoring this listing and other warnings from conservation organizations, Norway allowed the construction there of Europe's largest land-based wind-generating facility. It produces 450 million kilowatt-hours of electricity a year, but in doing so has killed 20 of the 21 local white-tailed eagles, according to the British Royal Society for the Protection of Birds (RSPB).

Nevertheless, like the U.S. Audubon Society, the British RSPB supports wind power. Speaking for that society, Dr. Mark Avery said, "The RSPB supports increased renewable-energy generation as part of a balanced approach towards tackling climate change, which we see as the greatest threat to the world's wildlife. However, we will object to any wind farms that seriously threaten important populations of birds and their habitats."

Both the U.S. Audubon Society and the British Royal Society for the Protection of Birds believe that the solution lies in adequate site assessment. They have created guidelines for wind-power installation that include several years of monitoring bird abundance, along with statewide planning so that wind turbines can be placed in locations favorable for both energy production and the birds. Their approach to wind energy is a model that we should keep in mind as we search for new ways to meet our energy needs. The best solution will be one that makes use of a variety of sources, each used to its best advantages.

The bottom line

- Wind-energy technology has matured rapidly in recent decades and is no longer simply experimental but a cost-effective energy source providing electricity today for distribution on the grid in the United States and other nations.

- Wind energy is playing an important role in rural areas and in poor nations, where other sources of energy are limited and expensive.

- Wind is cost-competitive with coal and is the least expensive of all alternative energy sources. For the cost of the Iraq War, the United States could have installed enough wind turbines to generate all its electrical energy needs.

- Environmental concerns about wind energy focus mainly on landscape beauty and bird mortality. The Audubon Society and other environmental organizations believe that these problems are greatly outweighed by its environmental benefits.

- Without question, wind will be a major part of the solution to our energy problem.

7

Solar power

FIGURE 7.1 Solar race car designed and built by students at the University of Minnesota passes by a wind farm near Lake Benton, Minnesota. *(DOE photo)*

Key facts

- Solar is the fastest-growing energy source.
- It presently provides a tiny fraction of the world's energy, considerably less than 1%, but if just 1% of Earth's land area had photoelectric devices, all the world's current energy needs would be met.

- And in about 20 years the solar energy generated on 1% of Earth's surface would equal the amount of energy in all known fossil-fuel reserves.
- The same is true for the United States—1% of the land area would provide *all* of the nation's energy needs, not just electricity.
- But at present, solar energy is the most expensive source of on-the-grid electricity, costing about twice the average price of electricity.
- However, solar energy can also be produced locally and in small amounts, and small solar-electric facilities are rapidly becoming more important in developing nations.
- In Kenya, some 10,000 women have been trained to use $10 solar cookers and show others how to use them. Hundreds of thousands of these solar cookers are also in use in India.
- In the 1990s, in Kenya alone, 120,000 photovoltaic units were sold to provide electricity for lighting, radio, television, and so on.

Crossing Australia at almost 60 miles an hour

Every other October since 1987, solar-powered cars have raced from Darwin to Adelaide, Australia, an 1,800-mile route that puts the latest alternative-energy technology to the test. The cars can run only on sunlight that their solar cells capture and convert to electricity. Electric motors that are at least 90% efficient are necessary. More than 50 teams—usually of college students, often backed by major aerospace and high-tech corporations—compete in the race, which takes a week.

In 2007, among the entrants was an all-women's team from Annesley College in Adelaide; the team designed, built, and raced their car. Only 18 cars finished. The winning 2007 team, from the Netherlands, completed the race in 33 hours at an average speed of 57 mph. A car from the United States, designed, built, and raced by students from the University of Michigan, made it in just under 45 hours at an average speed of 42 mph. The only other U.S. entrants did not even finish the race: The Equinox from Stanford University reached 1,158 miles, more than halfway, and Houston's Sundancer made it only 89 miles down the road from Darwin.[1] (Was this a sign that the United States might be losing its place as No. 1 in high-technology inventions, research, and development?)

As I write this in the summer of 2008, a website announces the next challenge in this race from the Swiss team, which promises more and more advances. Solar-powered car races aren't all that frequent, but devices powered by the sun are becoming familiar, even ordinary. In the United States, if you look carefully you are likely to find solar-electric devices near home. A drive along a major highway reveals emergency telephones and emergency highway signs powered by solar-electric panels. Some wristwatches are solar-powered, and you can buy a little solar electrical generator to recharge your iPod and your Blackberry.

At a much larger industrial scale, solar energy parks, large facilities that generate electricity that goes onto an electric grid, are rapidly increasing in number, and at this point it's hard to know at any time which is actually the world's biggest working facility (Figures 7.2 A, B). At this writing, the prize goes to two locations in Spain, both with a 20-megawatt capacity—one in Jumilla, Murcia, and the other in Beneixama. Both are said to provide enough electricity for 20,000 houses, and both are where you would expect to find solar energy installations—in a warm sunny climate. (These sites average 300 sunny days a year.)

FIGURE 7.2 (A) Largest solar photovoltaic installation in the Western Hemisphere, near Orlando, Florida. *(Photo courtesy of SunPower Corporation)*

FIGURE 7.2   (B) Obama at that Western Hemisphere's largest solar park. *(Photo courtesy of SunPower Corporation)*

The kinds of solar energy

Two kinds of solar energy are in use: active and passive. Passive solar energy is at work in buildings designed to absorb and reflect sunlight in ways that reduce the need to burn fuels for heating or cooling. No motors are used to move air or water or any material storing the energy. The way an automobile heats up when parked in the sun is an example of passive solar energy. Passive solar energy is often discussed as part of energy conservation, and we discuss it in Chapter 10, "Transporting Energy: The Grid, Hydrogen, Batteries, and More." Right now we'll talk about active solar energy.

Active solar energy uses electrical, electronic, or mechanical technologies to store, collect, and distribute solar energy for heating and cooling; to generate electricity; or (much more rarely) to do mechanical work directly. This chapter focuses on active solar energy technologies as sources of energy, primarily to generate electricity, but also, once the electricity is available, to use that energy to make chemical fuels.

Active solar energy, in turn, divides into two major technologies: those that convert sunlight directly into electricity, and those that use sunlight to heat (and boil) water, which is then used to run a conventional electrical generator or to provide hot water and space heating in a system that involves mechanical pumps and various control devices, including computers. (The pumps and controls distinguish active solar energy from passive solar heating of buildings.)

Solar thermal generators: sunlight to steam to electricity and big bright lights

The first large-scale test of using sunlight to heat water and using that to run an electric generator was Solar One, funded by the U.S. Department of Energy, built in 1981 by Southern California Edison and operated by that company along with the Los Angeles Department of Water & Power and the California Energy Commission. Here's the way it worked: Sunlight was focused and concentrated onto the top of a tower by 1,818 large mirrors (each about 20 feet in diameter) that were mechanically linked to each other and tracked the sun. It is a system that would have delighted Archimedes, who is said to have used mirrors to focus sunlight on enemy ships and burn them, probably the first military use of solar energy.

Solar One became famous to those of us in Southern California interested in the environment. I was teaching at the time at the University of California, Santa Barbara, and I drove on I-40 to Barstow to see this remarkable new device. It was impressive, especially when viewed in late afternoon among the long shadows of the surrounding desert. The mirrors reflected so much sunlight onto the tower that its top seemed aglow, as if it were a miniature sun emitting its own, not reflected, light. In fact, it was so bright that you couldn't look at it directly for long, as the accompanying photograph shows somewhat (Figure 7.3).

FIGURE 7.3 Solar One, the first major solar thermal tower, was built in the Mojave Desert near Barstow, California, several hundred miles east of Los Angeles. Originally it was an experiment funded by the Department of Energy and built and operated by Southern California Edison. *(Photo by Daniel B. Botkin)*

Solar One, still at the same location in the California desert, was improved in the mid-1990s and called Solar Two. Its mirrors and everything else necessary to run the system took up 126 acres—the mirrors alone had an area of 20 1/2 acres of reflecting surface!—and it operated with a capacity of 10 megawatts until it was shut down in 1999. Even today, it stands out in my memory as the most impressive view of the use of solar energy, almost magical in its brightness.

More recently, solar devices that heat a liquid and produce electricity from steam have used many mirrors without a tower, each mirror concentrating sunlight onto a pipe containing the liquid (Figure 7.4). This is a simpler system and has been considered cheaper and more reliable.

FIGURE 7.4 Researchers analyze the efficiency of using parabolic troughs for solar thermal systems. *(DOE photo)*

Solar electric generators: using very large, smooth surfaces to convert sunlight to electricity

The other major active solar energy technology uses solid-state devices that generate an electric current when exposed to light. If they weren't so common today, they would seem quite magical, like the solar energy flashlight I bought recently—you just lay it on the windowsill, and whenever you need light to use in the dark, you have it. These devices are called *photovoltaic cells,* and the most common of them is a silicon wafer, which is made of crystals of silicon. These make up 60% of the devices sold.[2] This is the same kind of material used to make computer integrated circuits. In fact, one photovoltaic company, Astropower, now out of business, used rejected materials from the manufacturing of computer chips. Silicon, by the way, combined with oxygen makes up most of the sand on beaches and is a very common material.

More recently, photovoltaic thin films have been developed, and some are in commercial production. (These make up almost 40% of the devices sold.) These use very small amounts of certain rare metal compounds, including cadmium telluride (a compound of cadmium and tellurium), CIGS (a chemical compound of copper, indium, gallium, and selenium), and a noncrystalline form of silicon.

At the time of this writing, some research is also exploring organic compounds that emit electricity when exposed to light, and there has been research over many years to try to take chlorophyll from green plants and use that in nonliving materials to generate electricity. But the majority of research and development of active solar energy is with inorganic photovoltaic systems.[3]

How much energy does solar provide now?

The amazing thing about commercially available photovoltaics is how efficient they are—how much of the energy they receive as sunlight goes out a wire as useful electricity. The record so far is 24% from crystalline silicon, and the thin films have reached 18% and 19%. Compare this to green plants that on average fix a miserly 3%.

By the end of 2007 there were many large installations, called *solar energy parks*. The 25 largest of these each had an electrical generating capacity of 5 megawatts or more from photovoltaic cells. Eleven of the largest installations were in Spain, ten in Germany, one each in Portugal and Japan, and only two in the United States.[4] These 25 largest solar parks by themselves have a generating capacity of more than 225 megawatts, enough to provide electricity for more than 225,000 homes. Worldwide, 880 photovoltaic power plants, each with at least 200-kilowatt capacity, are in operation, with a total capacity of 955 megawatts. Most of these large-scale plants are in Germany (390), the United States (225), and Spain (130).[5]

Although these individual systems are impressive both visually and in their electrical generation, they are dwarfed by the largest solar thermal plants. In fact, the largest solar energy installation of any kind operating today is Solar Energy Generating Systems (SEGS) in the Mojave Desert, the same region where Solar One was built. This is actually nine individual solar thermal plants with a combined capacity of 310 megawatts. Its 400,000 mirrors occupy 1,000 acres (about 1.6 square miles).[6]

Operated by Florida Power & Light and Southern California Edison, its capacity is larger than the sum of the 25 largest photovoltaic installations in the world.

What contribution do these systems make to the total energy supply? In the United States, total energy generated from all sources in 2005 was 29,590 kilowatt-hours.[7] (This is the standard value we are using throughout this book.) In 2008, the most recent year for which data are available, electricity generated in the U.S. totaled 4,119 billion kilowatt-hours. Solar energy provided just 0.02% (two-hundredths of a percent) of the electricity and 0.003% (three-thousandths) of the total energy from all sources. Wind provided much more: 1.34% of the electricity and 0.86% (almost 1%) of the total energy from all sources. So together, these two kinds of renewable energy provided a very small amount of the electricity and of the total energy. The situation is similar worldwide. At present, solar photovoltaic electricity provides less than 1% of the world's electrical energy and only about 0.1% of the world's energy.[8] So the take-home lesson here is that today solar photovoltaic systems provide a very small percentage of the world's electrical energy and therefore of the world's total energy use.

How much energy could solar energy provide?

The short answer: a huge amount. I got interested in this several years ago when what was then the largest photovoltaic system in the world, the one in Bavaria, began operating. I made calculations, based on *installed* facilities—not theoretical possibilities—for a variety of sites established by PowerLight Corporation (now part of U.S. SunPower Corp.). Here I need to make a disclaimer. My son, Jonathan Botkin, is an engineer at this company and has been especially helpful in making sure that my calculations are correct. When I gave a talk about solar energy to some journalists a few years ago and mentioned this, they seemed to think I was therefore lobbying and advertising for that corporation. Although I do think this company is doing a good job, I'm not touting them, just turning to someone I trust implicitly to make sure my figures are correct.

Here are some of the calculations I made. Although the largest solar energy parks are in Spain, it wasn't very long ago, in 2005, that the world's largest solar energy facility started producing electricity in what would seem an unlikely place, the small city of Mühlhausen, in Germany's Bavaria (Figure 7.5). Mühlhausen is famous for Johann Christian

Bach's life there as an organist, but although it's a picturesque tourist destination, it probably wouldn't go on your list of sunbathing resorts. Still, in 2005, one of the best answers to our energy crisis was right there, in farm fields where sheep grazed beneath an unusual crop: an array of rectangles mounted on long metal tubes that rotate slowly during the day, following the sun like mechanical sunflowers. Yes, this was, a few years ago, the world's largest solar electric installation, generating 10 megawatts on 62 acres. *Scaled up to just 3.5% of Germany's land area, this kind of solar power could provide all the energy used in Germany— by cars, trucks, trains, manufacturing, everything!*

Here are some U.S. examples. Based on already installed photo-voltaic systems in San Francisco's Bay Area, one acre covered by a photo-voltaic system provides enough electricity for 379 houses.[9] San Francisco covers 46 square miles and in 1990 had a population of 723,959. Figuring about three people per house, this is equivalent to 241,320 houses, and just 1,910 acres of solar collectors would be enough to provide all the domestic electricity for the residents. Thus, solar photovoltaic devices would occupy only 6.5% of the city's land area, and if the solar collectors were on house roofs, it is quite likely that the existing roofs would pro-vide adequate area for domestic electricity needs.

Let's use the same basic information for Arizona, even though it actually gets a lot more sunlight than San Francisco, but just to be on the conservative side. Arizona occupies 113,642 square miles of land, or 72,730,880 acres. At the installed energy yield I have been discussing— enough to provide electricity for 379 houses per acre—the entire state could provide electric power for 27 to 28 billion houses, and at an aver-age of 3 people per house, that's enough for 81 billion people, or 13 times the population of the Earth. If 1% of Arizona's land area were used for photovoltaics, enough electricity could be produced for more than 275 million houses, which is considerably more houses than exist in the United States. At an average of three people per house, this area of pho-tovoltaics would provide electricity for 837 million people, or about 28% of the world's population.

Other people have made similar calculations— for example, Profes-sor Nathan Lewis of Caltech, a expert on energy supply. He writes that in 1990 the total world energy demand was 10 billion kilowatts, and that using reasonable estimates of the increase in the human population and the increase in the demand for energy, by 2050 the demand could more than double, so that 28 billion kilowatts might be required to meet all energy demands.

FIGURE 7.5 The Bavaria Solarpark, in 2005 the world's largest solar energy installation, lies within a picturesque landscape famous in history but not especially sunny. *(Courtesy of SunPower Corporation)*

Professor Lewis also points out that our planet receives 120,000 billion kilowatts of energy continuously, on average, from the sun. So current world energy use by people is just two-hundredths of the total energy our planet receives from the sun.

This means that all the energy used by the world's people in 1990 could have been provided by covering just 0.1% of Earth's area (and just 5.5% of the United States) with photovoltaics that were just 10% efficient, and that the estimated energy demand in the year 2050 could be met by the same photovoltaics covering 0.16% of Earth's surface, or 8.8% of the United States.[10] Indeed, the total energy demand in the United States alone could be met if photovoltaics occupied 1.7% of the land. In about 20 years of collecting solar energy in this way, we would have collected as much energy as is contained in all known fossil-fuel reserves—and that's assuming a conservative efficiency of 10% in transferring solar energy to electrical output.

In short, the United States could become a net exporter of energy, either in the form of electricity or in the form of a fuel made with that electrical energy, such as hydrogen or a small hydrocarbon derived from hydrogen and carbon dioxide. And even if things did not work out exactly as suggested here, and required twice as much land area, we are still talking about a small portion of the land area of a nation and of the world.

Another approach: solar energy off the grid

There is an ongoing debate between proponents of on-the-grid and off-the-grid alternative energy. *On-the-grid* refers to solar energy whose electricity is put directly onto the electrical grid and becomes part of a major energy system for a region and for an entire nation, and in this way contributes to the world energy supply. *Off-the-grid* refers to solar energy whose use is local, ranging from providing power to a single house, to individual housing developments, to small villages, or small industries.

One can make the case that the on-the-grid/off-the-grid debate goes back to the very origins of the development of electrical energy during Thomas Edison's time. After inventing the lightbulb and helping with the invention and development of electric motors and generators, Edison promoted a direct-current system. But the problem was, direct current could not be transmitted efficiently over a long distance; there was just too much power loss. Edison lost out in the advancement of electricity when the first large hydroelectric plants were installed at Niagara Falls and the electricity was transmitted long distances as alternating current.

Later, the establishment of the Bonneville Power Administration and the Tennessee Valley Authority in the 1930s to build large hydroelectric dams enhanced and increased a regional, centralized electrical production and distribution system. It was viewed as government providing a service to all the people, but was also consistent with a large-scale, big-industrial approach to providing energy.

A different political philosophy with a different technological approach, could have taken the same money and promoted local production of electricity. But the technology wasn't really ready for that. Only now, with the development of late-20th-century and early-21st-century wind and solar energy technologies, can this kind of intense off-the-grid electrical generation be seen as a competitive approach to energy production.

Off-the-grid solar energy for rural areas, for the poor, for single-family homeowners, and for less-developed nations

Like wind power, solar energy is useful for those who lack easy access to other forms of energy or simply prefer to be energy-independent. The potential for off-the-grid, locally generated solar energy to help people in developing nations led to the development of solar cookers, but the earliest versions did not become widely used. These early solar stoves concentrated sunlight enough to cook food, but they were unfamiliar devices, out of context for the cultures and rural societies they were supposed to help, and have been referred to as solutions in search of a problem. It would probably be more accurate to say they were a Western male engineer's concept of a device to be used in non-Western civilizations by rural women used to cooking in traditional ways.

All that has changed rapidly. In recent years, the most famous proponent of solar cooking for developing nations is Margaret Owino, director of Solar Cookers International, East Africa, who has successfully promoted the use of these cookers in Southern Africa. More than 10,000 women in Kenya have learned how to use and promote them, and there are many testimonials about how they are helping. For example, Elizabeth Leshom, who lives in Kajiado about 50 miles south of Nairobi, has found that the solar cooker has cut her family's use of charcoal in half and considerably reduced her use of firewood.

Solar cookers come in two major types, simple *hot boxes* (Figure 7.6, top) and parabolic mirrors that concentrate sunlight onto a point (Figure 7.6, bottom). Hot boxes simply heat up enough to cook food. Their advantages are that a number of pots can fit into one box and they are cheap. The least expensive of these are made of waxed cardboard cartons with foil surfaces and can heat up to 275°F. In Kenya, these have sold for the equivalent of $5.60 to $7.90. Worldwide, these are the most widely used individual solar cookers, with several hundred thousand in use in India alone.[11]

FIGURE 7.6 (Top) A hot-box solar cooker *(Solar Cookers International)* and (Bottom) a parabolic solar cooker, whose mirror concentrates the heat at a point. *(Maarten Olthof/Vajra Foundation).*

Opportunities for entrepreneurs

According to an article in the *Wall Street Journal,* some companies are beginning to see a market in small and inexpensive solar energy devices for the Third World, including lighting. One company, Cosmos Ignite, sells solar-powered MightyLights for US$40, about the cost of a few months' supply of kerosene. This company came about as a project in a class at Stanford Business School, where the students were asked to develop a cheap alternative to artificial lighting for developing countries. One of the students, Matt Scott, developed a light and founded this company. In India alone, 10,000 of these have been sold.[12]

To give you a clear picture of the potential, here are some basic facts. In full sunlight, a square meter (think a little bigger than a square yard) receives 1,000 watts of sunlight, enough energy to light ten 100-watt lightbulbs. A silicon chip 4 inches square generates 1 watt of electricity in full sunlight at ordinary outdoor temperatures. A small photovoltaic system that generates 50 watts in full sun provides enough energy "for four or five small fluorescent bulbs, a radio, and a 15-inch black-and-white television set for up to 5 hours a day," according to Robert Foster of the Southwest Technology Development Institute at New Mexico State University.[13]

In Kenya, more than 200,000 small solar-electric systems have been sold. These include basic array of photovoltaic units, a storage battery, and whatever wiring and electronic devices were necessary to make a system work.[14] These have provided home lighting and power for radios, televisions, computers, and so on at a cost of a few hundred dollars for the smallest system (less than 16 watts output) to more than $800 for larger systems (more than 45 watts).[15] The cost to install these systems worked out to be $15 to $18 per watt, and of course there were no monthly fuel costs. By comparison, a gasoline electrical generator cost at least $500 to install and required $64 worth of fuel per month to run. Rural Kenyans without electrical generating systems buy kerosene for lighting and dry-cell batteries to operate radios, at a cost of $5 to $10 a month. Thus, a small photovoltaic system pays for itself in two years or so. Many small companies have started up in Kenya to sell, install, and maintain these systems.

By the end of the 20th century, only 62,000 Kenyan households— less than 1% of the total in that nation—were on an electrical grid.[16] A good argument can be made that in nations with economic situations similar to Kenya's, an electrical grid is not cost-effective and is unlikely to be developed; therefore, local, off-the-grid small systems will be the major way for their citizens to have access to computers, radio, television, and other modern technology. Small solar energy systems like those described above are helping.

For example, on the Philippine Island of Mindanao, U.S. AID (Agency for International Development) has funded the installation of solar cells (and small hydropower systems) to provide electricity where it was difficult to establish an electrical grid. I was a Peace Corps volunteer on that island in the 1960s. We lived in Marawi City, the capital of Mindanao Province, on the shores of beautiful Lake Lanao. About 30 miles down the river that flowed out from that lake, at the city of Iligan, was a hydroelectric power plant at Maria Christina Falls. The city of Iligan, on the coast and right near the falls, had a good electrical supply, but Marawi City had no grid system and no electricity except for individual gasoline and diesel generators. At the university where I was teaching, there were three engines: an electrical generator, a water pump, and a refrigerator. One day all three broke down, reminding all of us of the benefits of an electrical power system with some redundancies.

There were telephone lines that had been constructed when the Philippines was a U.S. territory, but with the renewal of fighting between people of Mindanao and the central Philippine government, the system had fallen into disrepair, and we were told that bandits had stolen the telephone wires to sell the copper. More recently, fighting between the Mindanao Muslims and the central government also made it impossible to develop an electrical grid. As a result, the Alliance for Mindanao Off-grid Renewable Energy (AMORE) began a new program that has electrified more than 500 villages with these small local systems.[17]

Other definitely off-the-grid solar technologies

Solar-powered vehicles. Paul MacCready, already famous for creating a human-powered aircraft that flew across the English Channel, designed and built *Solar Challenger*, which in 1981 flew from Paris to Canterbury, England, across the English Channel, flying a total of 163 miles and reaching an altitude of 11,000 feet. NASA developed solar-powered aircraft that have greatly exceeded that record.[18]

Space travel and solar energy. If you're planning a trip to outer space, perhaps to the moon or to Mars, you have two sources of energy for the long term, nuclear and solar, and space vehicles make use of both. The space station relies on solar energy, and so do the cute little Mars rovers that captured public attention when they began to rove slowly over the Martian landscape doing the bidding of Earthbound planetary geologists.

Downsides

Why aren't nations rushing even faster to install solar power facilities? And especially, why isn't the United States—the world's largest energy user—rushing to become a world leader in solar energy production? We can ask the same question about China and India, the two most populous nations and the two that have had the most rapid recent increase in energy use.

Costs

Today, primarily because of the cost of manufacturing photovoltaic devices, electricity from solar energy is more expensive than from most other sources, including fossil fuels. In the United States, according to the Department of Energy, electricity from solar energy costs 21¢ per kilowatt-hour for industrial production and 38¢ per kilowatt-hour for residential production,[19] while the national average price to consumers is 13¢ for industrial users and 16¢ for residential users.[20] Solar thermal towers and systems with parabolic reflecting mirrors have been cheaper to operate, providing electricity at about 12¢ per kilowatt-hour.[21]

Are the costs prohibitive? In 2002 Con Edison built New York City's largest commercial rooftop solar energy system for $900,000, providing energy for 100 houses. Assuming an average of three people per home, the installed cost is $3,000 per person. For all 300 million U.S. residents, the installation cost would be $900 billion.

The U.S. balance of trade is in the red by about $60 billion a month, or $720 billion a year, and much of this trade imbalance is due to the cost of foreign oil. So, for the equivalent of one year's trade imbalance, the United States could pay at least 80% of the cost of installing solar energy facilities for all domestic electrical consumption. The war in Iraq—justified, many say, in part to protect our petroleum sources—has cost an official federal allocation of more than $600 billion. And the Pentagon's acquisition budget reached $1.6 trillion in 2007.[22] In March 2008, a report by Nobel Prize-winning economist Joseph Stiglitz estimated that the true direct costs of the Iraq war will be $1.5 trillion or more, and the total costs, including the costs of health care and rehabilitation of veterans, will be more than $3 trillion.[23]

For the cost of the Iraq war, or perhaps just one-half or one-quarter of those costs, solar energy systems could have been installed to provide domestic electricity for all the people in America—energy forever!

As we saw earlier, the numbers become even more amazing for the dry, sunny climate of Arizona, where covering just 1% of the land with solar collectors would produce electricity for more houses than exist in the entire United States.

New solar cell technologies may lower costs. Although solar electrical devices are amazingly efficient, this technology is developing rapidly and costs could go down. Right now the best candidates for new kinds of solar collectors are thin film (a variation on the silicon cells that have been well established) and organic compounds. Crystalline silicon oxide, the material from which photovoltaic chips are presently made, is more expensive to make but more efficient than others. Amorphous silicon is used in "thin film" and is cheaper to manufacturer but less efficient. The engineering question is whether it is economically advantageous to pay more for higher efficiency of the fundamental receptor or to go with the cheaper basic unit. The latter will be the best choice only if the total cost is due largely to the cost of the primary photovoltaic cells.

This is not generally the case at present, but could change. Right now, photovoltaic cells represent about half the cost of a solar-electric installation,[24] suggesting that perhaps more costly, more efficient units are a better economic approach than cheaper, less efficient photovoltaic cells. But this debate is ongoing and will be resolved only by more research and development. It is beneficial at present to have both approaches taken, as is happening now because some corporations are producing the crystalline product and others are producing thin films.

Although silicon is the basis of most of today's photovoltaic cells, other chemical elements also produce a photoelectric effect, in particular cadmium and gallium-arsenide, toxic elements whose use should be either restricted or carefully monitored and controlled for health and safety.

Manufacturing limits

One of the major downsides, perhaps *the* major one, is that the manufacturing capacity to produce photovoltaics and solar thermal systems is presently inadequate to meet growing U.S. and global energy needs.[25] But the good news is that the number of photovoltaic cells manufactured in the United States is growing about 40% per year. If this rate of increase continues, solar photovoltaics could provide as much as one-third of the total energy the United States will need in 2050, as I discuss in Chapter 13, "Solutions." Whether this can happen without large-scale government investment that looks beyond the immediate market is unclear. Professor Nathan Lewis of Caltech writes: "Researching, developing, and commercializing carbon-free primary power technologies capable of 10–30 TW by the mid-21st century could require efforts, perhaps international, pursued with the urgency of the Manhattan Project or the Apollo Space Program."

Energy storage

A downside that is always pointed out is how to store the energy from sunlight. (This is true of wind energy as well.) The problem is perhaps most spectacularly illustrated by attempts to make solar-powered airplanes.

After Paul MacCready's successful design of the *Solar Challenger,* the solar-powered airplane that crossed the English Channel in the 1980s, NASA experimented with a remote-controlled solar-powered light aircraft called the *Pathfinder.* The best of these, *Pathfinder Plus,* flew to an altitude of 80,201 feet on August 6, 1998 (Figure 7.7). However impressive this was, the *Pathfinder Plus* had a limitation: It could carry only enough batteries for a few hours of flight after dark. As a kind of diurnal creature that had to land not too long after dark, it was not really a practical airplane.

There's been a lot of headshaking about the problem of storing the energy from sunlight and wind, as if this problem were unique to these two energy sources. But energy storage is also a problem for nuclear power plants, because they are most efficient when they run at the maximum electrical output all the time. In some cases, nuclear power plants have been linked to reservoirs, pumping water up into the reservoir at night when the demand for electricity was low, and generating electricity during peak demand from both nuclear reactions and waterpower. During a drought or a rainy season, hydroelectric dams and reservoirs, too, have a storage problem.

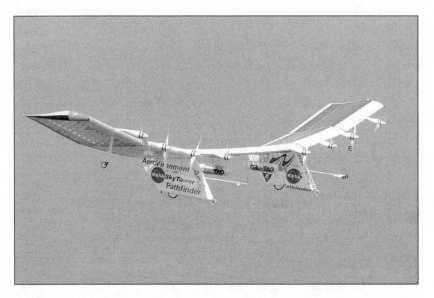

FIGURE 7.7   NASA's Pathfinder Plus solar-powered airplane.[26]
(Nick Galante/NASA Dryden Historical Aircraft Photo Collection)

We talk about solutions to energy storage in Chapter 10. In brief, the problem can be partially overcome by (1) connecting solar generators to the grid, (2) using solar energy to heat water, and (3) using the electricity to make gaseous and liquid fuels (starting with hydrogen taken from water). Of course, (4) storing electricity in batteries is always an option, but as NASA's *Pathfinder Plus* demonstrated, this has its limitations. We can also use the energy to do tasks for us whose timing is not very important, such as pumping water up into water towers for distribution later and desalinating water (processes that can be done whenever the energy is available).

Other means of storage have been proposed, and some tested. One of them is to store the energy mechanically in a flywheel and use that energy as needed by having the spinning wheel connected to an electric generator or directly to the wheels of a land vehicle.

Storage is not a simple problem with a single simple solution, but it is solvable, as I explain later.

Environmental effects: landscape beauty and competition for space

It would be naive to think that any source of energy had absolutely no undesirable effects, especially environmental effects. As Barry Commoner told us a long time ago, there is no such thing as a free lunch in nature. So probably some environmental problems will arise even from the use of solar energy. One that comes to mind is landscape beauty. Although solar collectors usually lie horizontally and thus have much less effect on scenery than do wind turbines, it will not be surprising if in some locations many acres covered by the black surfaces of photovoltaics are considered a blight on the landscape. As with wind power, solar facilities should be situated with the help of professional landscape architects and planners and experts in ecology to minimize potentially negative effects.

Solar parks will in some cases be seen as competing with other uses for land, but one advantage of some solar park designs is that the land can be open to multiple uses. For example, many solar installations are on rooftops. The most likely environmental negative of solar energy is with the mining, manufacturing, and recycling of materials, especially once solar becomes one of the world's major energy sources. Right now, recycling of batteries is not done efficiently. And while silicon forms some of the most common earth materials and is thus readily available, its mining creates fine dust that can be a local health and environmental problem and cause regional, even global, pollution if emitted high into the atmosphere. Surface mining for the large-scale manufacture of photovoltaic cells will damage landscapes and ecosystems in ways similar to surface mining, except that there will be less likelihood of acid drainage.

The bottom line

- The sun offers the greatest amount of energy, and could by itself, using a small percentage of Earth's surface area, provide the equivalent of all the energy used in the world by people.
- Solar energy has great potential and is benefiting from rapid increases in research and development, which will lower its costs.

- European nations are taking the lead in the installation of large solar facilities, with Germany and Spain outstanding users of this energy source.
- Solar energy is providing electricity and heat for cooking in many developing nations where a large-scale electrical grid does not exist and may never be practical.
- For many of the world's people, solar energy offers the only way to participate in modern, high-technology activities.
- Solar energy is bound to be a major player in the supply of energy in the future.

8

Ocean power

FIGURE 8.1 The *Suntory Mermaid II*—how to ride waves. *(Illustration by Kevin Hand, www.kevinhand.com)*

Key facts

- Energy in ocean currents, tides, and waves could provide twice the world's current energy use.
- Wave energy alone, using current technology, could provide 15% of the world's electricity.

- If the United States could harness 40% of the nearshore wave energy, it would capture as much energy as is generated by all freshwater hydropower now available in the U.S.
- The oldest use of ocean energy is from the tides. During their occupation of Great Britain, the Ancient Romans built a dam that captured tidal water and let it flow out through a waterwheel. In Medieval England, use of tidal power was not uncommon.
- The most successful modern tidal power plant is off the coast of Brittany, France, producing 10 million watts of electricity a year.
- Currently proposed is the world's largest tidal power plant, in Great Britain's Severn River estuary. It would have a generating capacity of 2 billion watts.

The wave of the future?

In March 2008, the *Suntory Mermaid II,* a new kind of boat, left Honolulu with a plan to travel more than 3,700 nautical miles (Figure 8.1). The trip had been done before, of course, but this boat's propulsion system was new—the *Suntory Mermaid II* has two horizontal fins that move up and down with the waves and generate the power to push the boat forward. Solar energy provides electricity. Dr. Yutaka Terao of the Department of Naval Architecture and Ocean Engineering at the Tokai University School of Marine Science and Technology invented the system that powers the boat.[1, 2]

Will this be the wave of the future, or just another futile attempt to harness the vast energy of ocean waves and currents and make it an important, practical alternative source of energy for us?

In Chapter 4, "Water Power," we talked only about power generated by freshwater—rivers, streams, hydroelectric dams. We are giving the biggie—the world's saltwater—a chapter all its own, not just because oceans are bigger but also because harnessing their energy is a whole different ballgame. The oceans are a vast renewable resource that, constantly in motion, could be a huge source of energy. The problem is that except in a few cases, their storms, currents, waves, and tides have been too powerful for our energy-converting machines.

Think of the ocean as a giant solar-energy collector, covering 70% of the Earth's surface. As the National Energy Research Laboratory explains it: "In an average day, 60 million square kilometers (23 million square miles) of tropical seas absorb an amount of solar radiation equal in heat content to about 250 billion barrels of oil. If less than one-tenth of one percent of this stored solar energy could be converted into electric power, it would supply more than 20 times the total amount of electricity consumed in the United States on any given day."[3]

The ocean holds two kinds of energy: one from its moving currents, waves, and tides, which we might informally call mechanical energy, and the other from the temperature difference between surface water and deep water, which we might informally call thermal energy. The World Energy Council writes that the ocean could provide the equivalent of "twice the world's electricity production,"[4] and that wave energy alone, with current technologies, could provide 15% of the world's electric energy.

These are rough estimates, to some extent limited to the efficiencies of existing technologies. It's hard to forecast improvements in efficiency and advances in the kinds of environments where ocean energy could be tapped, and therefore it is difficult to figure out how this energy could be captured and turned into electricity. Engineers generally try to take limitations into account. For example, another estimate assumes that only 20% of America's offshore wave energy would be harnessed, and that would be at 50% efficiency (meaning that half of the energy in the waves would end up as usable electricity). Even given these limitations, wave energy could still provide an amount of energy equal to all U.S. hydropower in 2003.

In other words, there's a hell of a lot of energy out there (Figure 8.2). The big trick is figuring out a way to turn that energy into reliable electricity with technologies that won't be quickly destroyed by ocean storms and the powerful eroding ability of seawater. There are a few successes, but big advances still lie in the future. The Electric Power Research Institute, a nonprofit backed by major U.S. power corporations, is sponsoring several projects to test technology for harnessing ocean energy.

Ocean motion

It is helpful to think of the mechanical energy in the ocean's moving waters as being of two kinds: tidal power along the shore and in river estuaries; and the power of offshore waves and currents.

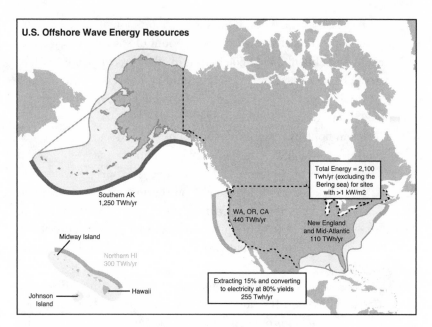

FIGURE 8.2 Some regions believed to have great ocean energy potential. *(Electric Power Research Institute)*

Tidal power

Given the difficulty of capturing ocean energy, it's surprising that the use of tidal power traces back at least to the Roman occupation of Britain. Archaeological excavations show that dams were built then to store water from high tides, and this water was released to run mills that ground grain.[5] In the Middle Ages, tidal power remained in use. The famous *Doomsday Book* of AD 1086 mentions tidal mills—more or less conventional water mills built to run off the tide as it flowed in and out.[6, 7] The Eling Mill in England, still in operation, is believed to date back to those times.

La Rance: proving that the tides can give us electricity for decades

One of the most successful modern tidal power plants is La Rance, along the coast of Brittany, France, which for years was the only full-scale tidal power station in the world. It was built in 1967 and produces 10 MW—enough to provide electricity for 300,000 homes, or for 4% of the population of Brittany—from 24 turbines, which produce 0.5 billion kilowatt-hours of power a year (Figure 8.3).[8] Wisely, this facility was built along a part of the Brittany coast that has one of the greatest tidal ranges in the world, about 40 feet between high and low tides. This maximizes the energy that can be obtained. Also impressive is that this power plant has never suffered serious damage or mechanical breakdown and, in contrast to many major energy sources that are eyesores, has become a tourist attraction.[9]

We can't expect La Rance imitators to be set up and work everywhere, because only a few places have such favorable topography. Another famous one is the Bay of Fundy in Canada, whose maximum tidal range is even a bit larger, at about 49 feet. There are also good sites along the northeastern United States. Present conventional technology requires a tidal range of at least 26 feet.

One disadvantage of a tidal-power dam is that it could restrict fish traveling up and down the river. There could also be other local damage to wildlife and fish habitats.[10]

FIGURE 8.3 La Rance Tidal Power Plant, said to be the only full-scale operational tidal plant of its kind in the world. It was built in 1967 and has continued to operate without suffering major damage from corrosive saltwater.[11]
(Dani 7C3/Wikimedia Commons)

Great Britain is currently proposing to build the world's largest tidal power plant in the Severn River estuary, the location of the world's second-largest tidal range, more than 45 feet. If built, it would produce 17 billion kilowatt-hours a year.[12] This is equivalent to a 2-million-kilowatt conventional power station. The British government proposes to spend $29 billion to build the facility, which means that the installation cost would be $0.07 per watt, cheaper than any other commercial form of energy. The tidal power plant is expected to run for 120 years.[13,14]

Environmental groups are opposing its construction, however. They say that it would threaten 70,000 acres of protected wetlands where almost 70,000 birds winter and also disrupt salmon, shad, lamprey, and sea trout migrations. They argue in addition that the project is too expensive and that there are other, much cheaper alternatives.

Proponents point out that if global warming raises the sea level as some climatologists forecast, the Severn estuary will be flooded anyway. And as for the cost, is it really so expensive?

Experimenting with ocean waves and currents

With all the energy potentially available from the ocean, a lot of imagination is going into designing machines to turn the energy in ocean currents and waves into electricity. The idea is to get away from building dams and other fixed structures, which fight the very motion that they try to use, and instead seek devices that would be immersed in the ocean and convert the motion of currents into electricity in a more forgiving way.

An Australian company has taken a bioengineering approach, using designs that occur in nature. One of their devices looks like a frond of giant kelp, the brown seaweed that forms underwater "forests." Like the kelp, the new device has holdfasts that anchor it to the bottom of the ocean and a flexible arm that waves back and forth with the motion of the water like the kelp's frond. This motion turns a generator that makes electricity.

Another device, shaped like a shark's fin, rolls with the moving water and lets extremely large waves pass by (Figure 8.4). Still another, called the Pelamis, is being developed by Ocean Power Delivery in Edinburgh, Scotland. The *New York Times* describes it as "a snakelike wave energy machine the size of a passenger train, which generates energy by absorbing waves as they undulate on the ocean surface."[15]

FIGURE 8.4 One of the new approaches to harnessing the ocean's energy is this device, shaped like a fish's fin so it can "swim" with the ocean currents and thus is less likely to be damaged. *(bioSTREAM™, Biopower Systems, ©www.biopowersystems.com)*[16]

Whether these turn out to be practical, efficient, and durable, only time will tell, but the potential energy is so great that this seems a worthwhile area to explore. In addition to the obvious benefits of not producing greenhouse gases, the devices are within the ocean, so they are not unsightly. One caveat: The extent to which they might affect ocean habitat, if installed on a large scale, remains unknown.

In addition to small start-up companies, some of the large electrical generator corporations are getting interested in this kind of ocean technology, including General Electric, the Norwegian company Norsk Hydro, and Eon Corporation of Germany.[17]

Thermal energy: using the ocean's temperature differences

The engines we are most familiar with today—gasoline and diesel internal combustion engines, steam turbines, and jet engines—are heat engines: They rely on the temperature difference between a hot and cold gas or liquid in different reservoirs to produce usable mechanical and electrical energy. Usually, the cool reservoir is the same temperature as the outside air or water, whereas the hot gas or liquid is heated by burning a fuel.

The idea of an ocean heat engine is not new. This classic kind of energy device was the focus of (and the motivation for) development of the science of thermodynamics in the 19th century. Jacques Arsene d'Arsonval, a French physicist, first proposed using the heat energy of the ocean in 1881. No one tried it, however, until 50 years later, when a student of his, Georges Claude, built an experimental system at Matanzas Bay, Cuba. It worked, producing 22 kilowatts—enough to light 220 100-watt incandescent lightbulbs. Five years later he built a bigger plant housed on a cargo ship off the coast of Brazil. The Cuban and Brazilian systems worked but were destroyed by storms.[18] More recently, in 1993, a system was installed by the Electric Power Research Institute at Keahole Point, Hawaii, and ran for a while as an experiment, during which it produced 50 kilowatts of electricity. These systems are still considered experimental and not ready for large-scale electrical generation.

In theory, you can produce mechanical and electrical energy using fluids of any two different temperatures, even if the temperatures differ by only a few degrees. But the closer the temperatures of the hot and cool fluids are to each other, the smaller the amount of useful energy yielded and the larger the machine must be to produce a practical amount. An automobile's internal combustion engine can be relatively compact because the burning gasoline within the cylinder is much hotter than the outside air. The pistons of an old-fashioned train's steam engine were also relatively small because the steam was so much hotter than the outside air.

In most places most of the time, surface ocean waters are somewhat warmer than deep ocean waters, and it was long ago realized that, at least in theory, you could operate a huge piston by pumping cold water from the depths and using it with the slightly warmer surface waters. However, one of the problems with developing a practical ocean-heat engine is that because there's only a few degrees' difference between deep and surface ocean waters, the engine would be huge and unwieldy, not likely to do well in storms.

The current focus is therefore much more on the technologies we discussed earlier that make use of "ocean motion." Still, given the huge energy potential of an ocean-heat engine, attempts to develop a practical one continue. According to the Electric Power Research Institute, "As long as the temperature between the warm surface water and the cold deep water differs by about 20°C (36°F), an OTEC [Ocean Thermal Energy Conversion] system can produce a significant amount of power. The oceans are thus a vast renewable resource." Such a system would provide about 10 billion KW, or two and a half times the amount of electricity available from present generating systems around the world.[19]

The bottom line

- The ocean offers huge amounts of energy from its currents, tides, and waves, but this is little developed, largely because of the problems posed by storms and the corrosive power of seawater. Progress seems possible, however, and this may be the best source of waterpower in the future.

- Tidal energy has been captured and used for several thousand years and is used today where conditions are best—a large tide and not much chance of severe storms. Considerable potential exists to expand tidal power in such specific locations.

- A proposed tidal power plant on the Severn River estuary in Great Britain would be the world's largest, providing 5% of that nation's electricity. It is, however, controversial among environmentalists because it could affect bird and fish habitats.

- Wave energy is obviously widespread, and rapid invention and technological development are under way. Whether the new devices will make tapping wave power practical enough to provide a significant amount of the world's energy is unknown, but it is promising enough to warrant research and development.

- The Electric Power Research Institute, a nonprofit backed by major U.S. power corporations, is sponsoring several projects to test technology for harnessing ocean energy. However, few large corporations and government agencies around the world have been set up to use ocean energy.

- Because ocean energy is still in the developmental stage, it is probably a place for venture capital and small start-up companies. But in a truly enlightened world devoted to renewable, sustainable, nonpolluting energy sources, funds now going to nuclear energy and the quest for more fossil fuels would be diverted to ocean energy technologies.

9

Biofuels

FIGURE 9.1 Harvest of experimental oilseed crops at Piedmont Biofuels farm, in Moncure, NC.[1,2] *(Photo by Debbie Roos, North Carolina Cooperative Extension)*

Interest in and enthusiasm about biofuels are growing among some small farms, like the members of the Piedmont Biofuels Cooperative in North Carolina. Will biofuel agriculture be the wave of the future and prove environmentally sound and sustainable?

Key facts

- The three kinds of biofuels are crops grown to be turned into fuels; organic wastes that can be burned for energy rather than dumped as is; and firewood.
- Biofuels currently provide only about 3% of America's energy, but two-thirds of this is from firewood and one-fifth from wastes. Agrifuels—crops grown to be made into fuels—provide only 12% of this total, or less than 1% of America's energy.
- Before the Industrial Revolution, firewood was a primary source of energy, along with the energy of domestic animals. At the turn of the 21st century, firewood was still important, providing 5% of the world's total energy use. In the United States, it provides heating for about 2 million homes.
- Firewood is especially important in poor, developing nations. In Rwanda, locally grown trees provide 84% of fuel for domestic cooking and heating.
- Wastes might provide as much as 8% of America's energy. Waste cooking oil makes a cheap fuel, but you have to be careful about its cleanliness, as dirt in it can clog internal combustion engines.
- Many scientific studies indicate that crops planted and harvested today to be turned into biofuels either take more energy than they yield or else provide only a small energy benefit.
- Any large increase in cropland devoted to biofuels will tax the supplies of water for irrigation and phosphate fertilizers. The price of such fertilizers has risen rapidly in the past several years because of increasing demand to meet the world's food needs.
- In the United States, biofuels are heavily subsidized. Direct subsidies are $1.00 a gallon for biodiesel produced as an agricultural crop directly for use as a fuel and $0.50 a gallon for biodiesel produced otherwise (such as from wastes).

Let me tell you about my father-in-law

My father-in-law, Heman Chase, a New Hampshire surveyor and country philosopher, heated his house with firewood he cut by himself for more than 30 years. He owned more than 200 acres of farm and woodland and kept about 60 acres of it as his primary woodlot. He also gathered wood from trees that blew down on his land or a neighbor's, who asked if he would mind clearing it away. Heman burned about ten cords of wood a year in a modern furnace that provided heat through a system of hot-water radiators. It was a lot of work to cut trees down, saw them into lengths that would fit in his pickup, drive them home, unload them, cut them into smaller chunks with a power rig he built for his small tractor, split the chunks by hand, and stack them. He enjoyed the work, but it did take a lot more of his time than he realized.

When he got into his seventies and wasn't well enough to do this any-more, he had a propane gas tank put in his backyard and a gas furnace installed next to the wood furnace. After a year he told me, laughing, that it was costing him so little to heat the house with gas that he probably should have saved himself a lot of time and effort years earlier. His ten cords of wood contained about the same amount of energy as 1,300 gal-lons of gasoline or 1,180 gallons of diesel fuel.[3] At early-1970s prices,[4] Heman might have paid about $600 for fossil fuel to heat his house—pretty cheap compared with what his time was worth as a professional surveyor and how much time he spent on his firewood. I don't think he would have made that choice, but it tells us something about what is required if you want to go the biofuels route.

Heman built his house with his own hands in the 1930s. Most of us think nobody back then gave much thought to the environment, but he and his New England friends were conscious of their surroundings and tried to care for them. Even so, Heman's 60-acre woodlot looked quite different from his other woods. The species were different, because he cut out the trees that made the best firewood—sugar maple and oak—and left what he considered trash trees, like poplar. The woodlot was more open, and the soil looked more compressed and used.

Use of wood and dung as fuel today

My father-in-law had a choice—many people today don't. Before the 20th century, wood was a major fuel in the United States. Many people in the world still use it. Around the turn of the 21st century, 5% of the world's total energy use was supplied by about 1.5 billion tons of fuel-wood.[5] People who live in developing nations continue to depend on firewood and animal dung as their primary fuels for heating and cooking. In Rwanda, for example, locally grown trees provide 84% of domestic cooking and heating fuel. At this point, consumption outstrips the supply, so Rwandans are facing a firewood shortage.[6]

As arduous as Heman Chase's use of firewood in New Hampshire might seem, still harder is the collection of firewood for cooking among the Maasai in Kenya. The grueling task falls to women and girls, who must walk long distances to collect wood and then carry their heavy loads of wood home on foot. On average, by the time a Maasai girl reaches the age of 16 she has carried 16 tons of wood home, on each trip carrying half to two-thirds of her own weight.

Perhaps more than any other story in this chapter, the toil of the Maasai and other African women to collect firewood by hand points out how important energy is for human life, and how difficult it can be to obtain the material goods necessary for life without modern machines and superabundant energy.[7]

Since the beginning of the 20th century, the use of wood for home heating has gone in and out of fashion several times in the United States. With easy access to inexpensive fossil fuels, a general transition away from firewood took place, especially in cities, by the mid-20th century, although it continued to be used widely in rural areas. Then in the 1960s, the era when hippies and flower children and many others began a back-to-the-land migration, woodstoves became popular again. So popular that small valleys in Vermont, New Hampshire, Colorado, and California suffered air pollution on days of temperature inversions, when the wood smoke from all those homes couldn't get out over the mountains. As a result, the Environmental Protection Agency established regulations for clean-burning woodstoves, and firewood went out of fashion. As recently as 1993, some 3.1 million homes were heated with wood; but by 2001 the number had fallen to only 2 million.

Recently, with the price of petroleum skyrocketing, woodstoves became popular again. By midwinter of 2008, New Englanders were heading back to stores to buy woodstoves. In an article in the *New York Times*, Roy L'Esperance, the owner of the Chimney Sweep in Shelburne, Vermont, said he has seen sales of woodstoves increase 20%. "There's a lot of people buying big stoves, planning on tackling oil head-on," he explained.[8]

Right now there are an estimated 40 to 45 million wood-burning appliances in the U.S., 15 million of them woodstoves. Only one-quarter of these were built after the EPA set up air-pollution standards for wood-stoves.[9] These contribute 430,000 tons—about 6%—of the total particulate-matter pollution in the U.S.[10]

Interest in biofuels has been growing

Biofuel (also called biomass power) became popular with environmental groups and agricultural industrial corporations, and the darling of Washington politicians, in the 2000s, promoted by agricultural subsidies. *Scientific American* liked it—a 2006 special issue featuring "Energy's Future Beyond Carbon" listed "15 ways to make a wedge" in carbon production. Number 13 was "drive 2 billion cars on ethanol, using one sixth of the world cropland."[11] As recently as May 5, 2007, an editorial in the *New York Times* supported corn-based ethanol.[12] Signaling this popularity, on February 24, 2008, Virgin Atlantic Airline flew a Boeing 747 from London to Amsterdam using 20% biofuel and 80% conventional jet fuel, the first test of biofuel by a commercial jet.[13] In the fall of 2008, Air New Zealand did a test flight with the same model airplane, fueling one engine with a 50-50 mixture of conventional fuel and biofuel. And at the time of this writing, the U.S. Air Force is planning in the spring of 2010 to fly an F/A–28 Super Hornet with a variety of biofuels.

How much energy do biofuels provide today?

Biofuels currently provide only about 3% of America's energy (Figure 9.2). Woodburning makes up 67% of this total (this includes wood wastes such as bark and otherwise unusable wood burned to provide heat in the production of paper); wastes 21%, and agrifuels only 12%. But ethanol from crops is increasing. In the United States in 2004, ethanol replaced approximately 2% of all gasoline (1.3% of its energy content). By 2007, it replaced 2.8%.[14,15] Worldwide, crops currently provide 13.5 billion gallons of ethanol a year.[16] Biofuel production has increased rapidly in the nations of the European Union, reaching 8 million tonnes in 2008.[17]

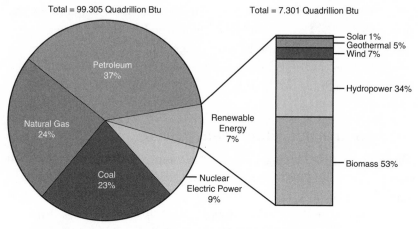

Note: Sum of components may not equal 100% due to independent rounding.

FIGURE 9.2   Biofuel use in the United States as of 2008. *(Source: Energy Information Administration, Renewable Energy Consumption and Preliminary Statistics 2008)*[18]

Fuel from waste

Atlanta-based Biomass Gas & Electric LLC, a developer of biomass-driven power plants, recently signed a 20-year agreement to supply power to Progress Energy Florida, a subsidiary of Progress Energy Inc., based in Raleigh, North Carolina. The deal involves a wastewood-to-energy plant not yet built that is expected to produce 42 megawatts. "We joke here in the office that [the Southeast] is the OPEC or Saudi Arabia of biomass," says Biomass CEO Glenn Farris.[19]

With increasing frequency, waste used to produce biofuel is from agricultural products that were grown and used for food or some other purpose and were then simply dumped. As a fuel, these serve a dual purpose: We don't need to set aside land to grow them since they've already been farmed for food, and we don't need to find places to dump them.

Take cooking oil, for example. An Asian restaurant named Pearl East, in Manhasset, Long Island, New York, used to pay $40 a month to get rid of its old cooking grease from making egg rolls and fried noodles.[20] But starting in December 2007, as part of a new environmental initiative by the Great Neck Water Pollution Control District, the restaurant's used grease and oil have been collected in 55-gallon drums and sent, at no charge to the restaurant, to the district's plant, where it is converted into biodiesel fuel and used to power the district's vehicles and generators. Not only does this get rid of a nasty waste, but also the fuel is much cheaper for the government, whose costs to transport and process it work out to only 70¢ a gallon, versus more than $3 a gallon to purchase diesel fuel at the commercial pump. Keep in mind that this wasn't grown to be used as fuel. It was grown for food and is now a waste product that can be used as fuel.

An interesting and amusing story by Susan Saulny in the *New York Times* on May 30, 2008, told of a "bandit" who "pulled his truck to the back of a Burger King in Northern California one afternoon" and "dunked a tube into a smelly storage bin and…vacuumed out about 300 gallons of grease." The police found 2,500 gallons of used fryer grease in his truck.

The article goes on to say that the owners of the Olympia Pizza and Pasta Restaurant in Arlington, Washington, claim their 50-gallon grease barrel has been siphoned several times, and they are considering installing a surveillance camera. "Fryer grease has become gold," an owner said, "and just over a year ago, I had to pay someone to take it away." Used frying oil is now traded on the commodities market as "yellow grease."[21]

In some developing nations, garbage can be a strong energy alternative. Masada Resource Group LLC, a privately held Birmingham, Alabama, developer of waste-to-energy plants, says it is planning to raise $60 million to help develop and operate commercial-scale plants in Central and South America.[22]

What crops are grown today to provide biofuels?

Corn and sugarcane are producing the largest amounts of biofuels at present, but many other crops are proposed, some have been tested, and quite a variety are in production. These include standard crops such as cassava, coconut, potatoes, sugar beets, sorghum, soybeans, and wheat; oilseed crops such as sunflower, mustard, and rapeseed (especially popular in Europe to make biofuels); pistachios (becoming popular as a biofuel source in China); alfalfa (a legume); and grasses, including reed canary grass and four natural grasses of the American prairie: prairie cordgrass, switchgrass, Indiangrass, and big bluestem. Some trees, in addition to those grown for firewood, are also in production for industrial biofuels, especially hybrids of poplar (one of the fastest-growing trees, grown extensively in plantations in Europe, where it provides much of the lumber and paper).[23] Mixes of natural or naturalistic vegetation are also proposed, such as American prairie plants.[24] These crops-for-energy have two major products: vegetable oils, like cooking oil, that can be used directly as fuel; and ethanol, good old mountain dew drinkable alcohol (ethyl alcohol, scientifically, not to be confused with methyl alcohol, the poisonous "wood alcohol") that has to be made with a distillery, as the people at Johnny Walker can tell you.

Fuel versus food

Biofuels may sound like the perfect solution to our fuel problems, but it is not—far from it. Consider the case of farmer Alfred Smith of Garland, North Carolina, who began feeding his pigs trail mix, banana chips, yogurt-covered raisins, dried papaya, and cashews, according to an article in the *Wall Street Journal*.[25] The pigs were on this diet, Mr. Smith said, because the demand for the biofuel ethanol, produced from corn and other crops, had driven up prices of feed (the largest cost of raising livestock) to the point where it was cheaper to feed his animals our snack food. In 2007 he bought enough trail mix to feed 5,000 hogs, saving $40,000. Other farmers in the U.S. Midwest were feeding their pigs and cattle cookies, licorice, cheese curls, candy bars, french fries, frosted Miniwheats, and Reese's Peanut-Butter Cups. According to the *Journal,* "Cattle ranchers in spud-rich Idaho were buying truckloads of uncooked french fries, Tater Tots and hash browns." Near Hershey, Pennsylvania, farmers were getting waste cocoa and candy trimmings from the Hershey Company and feeding it to their cattle.

Their problem was caused by competition with crops grown directly to be turned into fuels. We'll call these *agrifuels.*[26]

Proponents of agrifuels will search for more and more efficient methods to produce them, requiring less area and lower costs. But with agrifuels, Mother Nature sets an upper limit, because of the laws of physics and the limits to how much energy can be produced by green plants, algae, and bacteria. Growing plants to turn into fuel, we might be able to approach the upper limits of the efficiency of Mother Nature's photosynthesis, but we can't do better than the laws of thermodynamics allow, nor better than those amazing and intricate devices inside the cells of green plants can do—the photosynthetic machinery of life. For this reason, we have to ask whether Mother Nature can compete well with approaches completely different from crops-for-fuel, in terms of both the area required and the energy we get back versus what we put in.

According to David Pimentel, one of the world's experts on the ecology of agriculture and on biofuels, if the total U.S. production of corn were used to produce biofuels rather than food, the ethanol produced would provide only 5% of today's total oil consumption— 2% of the total energy used—by the nation.[27] Because corn doesn't produce ethanol directly, as any home distiller can tell you, the corn has to be fermented and distilled, which requires considerable energy. As currently carried out, producing ethanol from corn takes 46% more energy than is contained in the resulting fuel. In other words, making fuel from corn takes energy; it's not a source of energy. The same is true for ethanol from switchgrass, one of the native American grassland species that is being grown for biofuel, and rapeseed (the source of Canola oil). Ethanol produced from switchgrass or rapeseed is slightly less an energy sink than corn ethanol but still results in an energy loss rather than an energy source. The only benefit of the process is money made by agricultural corporations who received large subsidies for it.

Others propose using America's native grasslands to produce ethanol for fuel, but much of these are used for grazing, and the rest is important to biological conservation, soil conservation, and recreation. Still others suggest that agricultural "wastes"—stems, roots, and leaves that are not part of a harvested crop—should be used to make biofuels. But these organic materials are necessary for soil structure and fertility, to have productive croplands in the future, so this would be an unwise path to follow.[28]

Elsewhere in the world, large areas of tropical rain forests have been converted to palm oil plantations to produce biofuels. In contrast to the crop plants and grasses of North America, palm oil appears to provide a net energy output of about 30%. But these plantations are replacing valuable tropical rain forests and threatening species such as orangutans.

Basic considerations in judging the value of biofuels

Whether, on balance, biofuels are good or bad depends on (1) which biofuel—biological waste, wood for household use, or agrifuels from wide-scale industrial farming; (2) how it is produced; (3) what that production competes with; and (4) how energy-efficient and cost-efficient the production process is.

Judging a fuel's efficiency

Judging any fuel's efficiency involves four factors: energy efficiency, area efficiency, cost efficiency, and carbon efficiency. We discuss the first three. The last needs a book of its own and goes beyond what we can discuss here; this is not to deny its importance.

Energy efficiency is how much energy you get from a fuel compared with the amount of energy it took to produce that fuel. For a fuel to be useful at all, the efficiency has to be greater than 1 (the breakeven point, where it takes just as much energy to produce the fuel as you get out of it), and it should be a lot larger.

Area efficiency is how much energy you get from each acre you use to obtain that energy. With biofuels, it is how many acres it takes to get the equivalent of a barrel of oil or gasoline from, say, corn or sugarcane. Does one crop use a lot less acreage than another? If so, then there are benefits to using the most area-efficient crop, because this leaves more land to grow food, or for biological conservation, or for housing, and so on.

Cost efficiency is how much money you can get for a fuel versus how much it cost to produce it. This, too, should be greater than 1 if anybody is going to make money on it. But there's an important caveat: In certain situations the cost efficiency may be less than 1 but the fuel is still worth it. For example, if you are sending a rocket to Mars and want to power the landing vehicle with a small nuclear generator, you will be willing to expend energy to develop and build and transport that generator even though that energy will never be returned, because the goal is not to produce energy but to acquire information. In fact, the cost efficiency of all current forms of space travel is zero. Other examples in which cost efficiency is secondary are the use of nasty wastes like old cooking grease for fuel rather than dumping them into the environment, and backpacking fuel for warmth and cooking on a camping trip—hauling it along may be a lot of work, but it's worth it.

The energy efficiency of biofuels is questionable

For agrifuels—plants grown primarily for fuel—the exact energy efficiency is hard to determine because so many species are being used, tested, or proposed right now, and because accounting for all the energy used to produce a crop is complicated.

For example, energy is used not only directly by the farmer to grow the crop, but also to produce the seeds he uses, transport those seeds to the farm, and transport water, fertilizers, and pesticides that the crop requires. Manufacturing farm machinery and transporting it to the farm also take energy, as do constructing farm buildings and manufacturing and transporting the materials that go into them. It takes energy to feed, clothe, and house the farmer and his employees and to transport them to and from the cropland. It takes energy to harvest the crop, transport it to where it will be converted into a biofuel, carry out that conversion, and then transport the biofuel to locations where it will be used.

To make it even more complicated, some scientists count only the energy content of the biofuel itself on the positive side since this is the goal of the entire process. Other scientists add to that the energy content of all the remaining parts of the crop, including organic wastes and by-products, whether or not they end up being used as fuel. Thus, one careful analysis concludes that it takes 29% more energy to produce ethanol from corn than is contained in the ethanol.[29] Another concludes that there is a 25% gain, but the gain includes the energy in animal feed, a by-product, and in this second case the scientists conclude that the environmental and social costs outweigh the benefits.[30]

Studies of biodiesel fuels (where the crops yield vegetable oil that is then burned as fuel), too, differ widely as to whether they provide a net energy benefit. According to one analysis, biodiesel fuel produced from soybeans required 27% more energy than it yielded. Another study concluded that biodiesel produced "93% more usable energy than the fossil energy needed for its production,"[31] but, again, the more optimistic estimates usually include energy that could be obtained from parts of the crop that are not intended to be used as fuel and generally aren't.

The bottom line on biofuel energy efficiency

Many scientific studies conducted so far indicate that crops planted and harvested today to be turned into biofuels either take more energy than they yield or else provide only a small energy benefit. But there is a large range in the estimates, from a net energy benefit of 12 kilowatt-hours in a gallon of ethanol to a loss of 7 kilowatt-hours from each gallon produced.[32] The difference seems to depend mainly on how one calculates the energy costs to produce the gallon of ethanol, not on the actual energy content of that gallon. Even when they do appear to provide a net energy benefit, the benefit is outweighed by environmental and economic consequences (such as erosion, food scarcity, and higher food prices).

Proponents of biofuels argue that we don't need to be concerned about net energy benefit; they say it just doesn't matter. Instead, we need to compare only the energy content of biofuels with other fuels and the uses we can put each to. So, ethanol and biodiesel are better than coal simply because they can be used to power cars, trucks, and airplanes. But this argument requires that you suspend belief in the laws of thermodynamics, that you don't care that to get corn ethanol or vegetable oils you burn more fossil fuels than they yield. The argument is absurd, and I have mentioned it only because this is likely to come up in discussions of this chapter. So let me make the point clear: If a chemical takes more energy to make than it yields, it is not actually an energy source. It may be useful as a product, or, as I've mentioned before, when one goes on a long hike and wants to carry fuel for cooking and warmth, or on a space trip.[33]

The U.S. Energy Independence and Security Act (passed by Congress in 2007) calls for 38 billion gallons of ethanol to be produced from biofuels per year by 2022, an amount that equals a quarter of all gasoline consumed in the U.S. today.[34] And no more than 40% of this can be provided by corn. The rest has to come from crops other than small grains. Even under the best possible conditions of weather, soil fertility, irrigation, and lack of crop diseases and pests, this would require at least 118 million acres, more than one-third of the U.S. land in crops in 2009, so this requirement will compete heavily with food production (Figure 9.3).[35, 36]

Crops grown to produce biofuels will tax the supply of water and phosphate fertilizers.[37] There is already worldwide concern about the availability of freshwater, and crops grown to produce biofuels will place an additional burden on water supplies. Industrial agriculture makes use of large quantities of fertilizers, two of whose main ingredients are nitrogen and phosphorus. Industrial processes enable us to convert nitrogen gas in the atmosphere to nitrogen compounds for fertilizers, but phosphate rock is obtained from mines, and there are limits to its economic availability. Like fossil fuels, phosphate rock is distributed nonuniformly around the world. About 80% of phosphorus is produced in four countries—the United States, China, South Africa, and Morocco[38, 39]—and the global supply that can be extracted economically is estimated at about 15 billion tons (15,000 million tons).

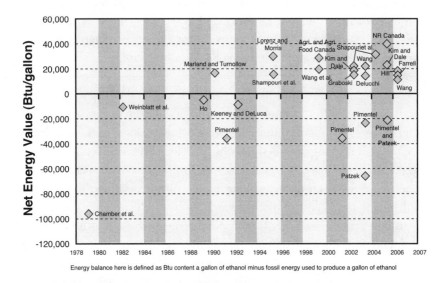

Energy balance here is defined as Btu content a gallon of ethanol minus fossil energy used to produce a gallon of ethanol

FIGURE 9.3 How much energy does corn provide? Studies vary widely. Most recent studies show a positive net energy balance for corn ethanol. *(Courtesy of Michael Wang, Center for Transportation Research, Argonne National Laboratory, personal communication)*

In 2009, the United States obtained approximately 30.9 million tons of usable phosphate rock from mines, 85% from Florida and North Carolina, the rest from Utah and Idaho.[40] In short, all of our industrialized agriculture—most of the food produced in the United States—depends on phosphorus from just four states!

Total U.S. reserves of phosphate rock are estimated at 1.2 billion metric tons. However, obtaining the 30.9 million tons of marketable phosphate rock from U.S. mines in 2009 required removing 120 million tons of rock from the mines, and in the next few decades phosphorus is likely to become even more difficult to obtain.

According to the U.S. Geological Survey, in 2007 the price of phosphate rock "jumped dramatically worldwide owing to increased agricultural demand and tight supplies of phosphate rock." And by 2009 "the average U.S. price was more than double that of 2007," reaching as much as $500 a ton in some parts of the world.[41]

One fact is clear: Without phosphorus, we cannot produce food. Declining phosphorus will harm the global food supply, thus affecting all our economies. Extraction continues to increase as the human population increases the demand for food, and as we grow more corn for biofuel production. To account completely for the costs of biofuels from crops, you would have to take into consideration its effect of the price of phosphate.

Mining, of course, may have negative effects on the land and ecosystems. For example, in some phosphorus mines, huge pits and waste ponds have scarred the landscape, damaging biologic and hydrologic resources. Balancing the need for phosphorus with the adverse environmental impacts of mining is a major environmental issue. The United States requires open-pit phosphate mines to be reclaimed to pastureland, an additional cost of biofuel production.

Biofuel proponents say that agrifuels can be grown on nonagricultural land—which is to say, land too rough and steep for farming. The trouble is, a lot of this kind of land is now used for recreation—such as hiking, climbing, skiing, and camping—and to provide habitat for wildlife and endangered species, natural ecosystems, and watersheds for streams and rivers that help sustain biodiversity. Thus, using these nonagricultural lands to produce biofuels can only be detrimental to biodiversity. What's more, land unsuitable for farming edible crops is also unsuitable for farming the same crops for conversion into fuels. Farming steep slopes leads to erosion; and high elevations and poor soil are too unproductive to be economically useful.

Biofuels from lakes and the seas?

It is difficult to make liquid fuels from land plants. First of all, the plants themselves do not produce ethanol directly (or if they do, it's in very small quantities). Second, a lot of the plant material is not directly convertible. Third, in most cases a lot of energy is spent transporting things to the plants (fertilizers, pesticides, water, farming equipment) and transporting the crop to a factory where the fuel is made. In sum, it is clear that taking cropland away from food production is a bad idea. Thinking back to solar energy, we realize that today's off-the-shelf photovoltaics are much more efficient than any land plant in making energy available to us.

These reasons are leading people enthusiastic about biofuels to turn to microorganisms, primarily aquatic algae and photosynthetic bacteria, but also to some microorganisms that are just plain good at converting woody tissue—cellulose in particular—to ethanol. The benefits seem great. These microorganisms produce a lot of biodiesel (oils that, like household vegetable oils, can be used directly in diesel engines) or ethanol directly, so we don't have to expend energy, materials, and time in making the liquid fuel. They can also grow in habitats that do not directly compete with food production. Where waters are polluted by fertilizer and sewage runoff and therefore are high in nitrogen and phosphorus, the algae and photosynthetic bacteria, which now create environmental messes for us, could play the double role of cleaning up the waters and giving us liquid fuels for aircraft, cars, trucks, and ships. This is all the more important because solar and wind produce electricity, which will play a major role in powering increasing numbers of electric cars, trucks, and rail transportation but cannot be used for air travel unless we use it to make a gas or liquid fuel (more about that in Chapter 10, "Transporting Energy: The Grid, Hydrogen, Batteries, and More").

Biofuels from microorganisms seems likely to give a large energy benefit, rather than being an energy sink or a net energy nothing as is the case with most of the land devoted today to biofuel production from land plants. David Pimentel, the expert on the ecology of agriculture and environmental effects of biofuels, tells me that ethanol from algae appears able to yield 50% more energy than it takes to produce it.[42]

SunEco Energy, in Chino, California, says that the algae in its 200 acres of aquaculture ponds yield ethanol with an energy content five times higher than the amount of energy used to grow the algae and process it—and the algae is always there. The company says it can produce ethanol for $0.81 a gallon, making it both energy- and cost-competitive, and projects that it can produce 33,000 gallons of what it calls "biocrude" per acre of ponds one foot deep.[43] Some of the major oil companies are getting into this game as well. Exxon plans to invest $600 million in an algae project.

One of the most interesting tests of producing algae from biofuels is a start-up involving the Southern Ute Indian Tribe in southwestern Colorado. They are funding and cooperating with Colorado State University Mechanical Engineering Professor Bryan Willson, who founded Solix Fuels. The company has an unusual approach—the algae are grown in bags in water tanks. The idea is to increase yield, but the production requires more energy and material input than other facilities.[44]

Another interesting development is the discovery of a bacterium that lives in forest soils and decomposes cellulose from the woody tissues of plants, giving off ethanol as one of its major products. The bacterium is known scientifically as *Clostridium phytofermentans* and more informally as the "Q Microbe," because it was discovered near the Quabbin Reservoir by Professor Susan Leschine and her colleagues of the University of Massachusetts. They point out that cellulose is the most abundant organic material on Earth. And, of course, it is in many of our wastes—papers, plastics. The Q Microbe has decomposed wood and sugarcane wastes, and Professor Leschine has set up a company, Qteros, which is doing research and development to use this microbe to both decompose wastes and provide ethanol fuel. Although still in its early stages, the work seems promising; once again, as with algae, this bacterium produces a fuel directly and therefore is likely to give a net energy return.[45]

The bottom line is that biofuels from microorganisms appear to be the most promising of the biofuels derived from organisms grown directly to produce fuel. But a lot of research and development are needed to find out just how productive of energy these can be. Right now, algae production of biodiesel seems to be the best bet for direct biological production of fuels, from both an energy-efficient and environmental point of view.

Are biofuels the answer?

I was recently on a business trip to Hailsham in southern England, about 100 miles south of London, and took a taxi back to Heathrow Airport. We drove through countryside, pastures full of sheep and this year's lambs, fields glowing bright yellow with this year's crop of rapeseed. The driver and I began talking about the local agriculture.

"That's the stuff I've been running this Mercedes diesel on for several years," he told me—"oil from rapeseed." I thought he meant that the crop was planted with the purpose of converting it into a fuel, but he said no, he'd just been going to the local supermarket and buying bottles of rapeseed cooking oil (marketed in North America as canola oil) and pouring it into the tank. He said he'd been a car mechanic and had read about this in a car magazine. He used new oil because he'd been warned that dirt and other contaminants in old cooking oil might damage his engine. When he took his diesel in for its annual inspection, the mechanic commented on how clean the engine was. "I don't notice any difference in the power or acceleration," he told me.

When Rudolf Diesel invented his engine in the late 19th century, he ran it on vegetable oil and thought that would be the fuel for it. This taxi was proof that vegetable oil did work well as a fuel for a diesel engine. So is this one of the best solutions for us?

Biofuels' effects on the environment

Biofuels can have various negative effects on the environment.

Destruction of soil and water

Let's start with ethanol from sugarcane in Brazil because that seems to have the best energy efficiency, reportedly between 1.7 and 9. (The low estimate means that the ethanol contained more than 1.7 times as much energy as was used to obtain it from sugarcane, while the high estimate is 1.9 times the energy input.)[46,47] The wide range of energy efficiency has to do with estimates of the energy used in transportation—of fertilizers and pesticides to the cropland, sugar to the mills.

Unfortunately, sugarcane is notorious the world over for being one of the crops most destructive to soil and water, especially polluting water runoff with soil particles, nitrates, and phosphorus, causing many problems downstream. Farming sugarcane erodes soil at more than five times the rate at which soil is being formed naturally in Brazil. It also takes huge amounts of water, especially to wash away soil that clings to the sugarcane. Washing each ton of sugarcane takes 1,900–9,500 gallons of water. Each acre of sugarcane also uses 59 pounds of nitrogen, 47 pounds of phosphorus, half a pound of insecticides, and 2.7 pounds of herbicides.[48]

Debate over biofuels and greenhouse gas

One of the issues raised all the time about global warming is that it might increase the rate of extinction of species, and that therefore we must deal with global warming directly.[49] This has become a major justification for biofuels. But the result is ironic: Land that is habitat for many species, including endangered ones, is being converted to profit-making agrifuel crops with the justification that this is helping the fight against global warming.

For example, in Asia, natural tropical forests are being cut and turned into coconut plantations to produce biofuel from palm oil. The net effect is to increase carbon released into the atmosphere (because deforestation leads to much decomposition of dead plants and organic matter in soils), to increase soil erosion and water pollution, and to destroy habitats for endangered species such as orangutans.[50] The environmental organization Friends of the Earth says that as much as 8% of the world's annual CO_2 emissions can be attributed to draining and deforesting peatlands in Southeast Asia to create palm plantations. The organization estimates that in Indonesia alone 44 million acres have been cleared for these plantations, an area equal to more than 10% of all the cropland in the United States, as large as Oklahoma and larger than Florida.[51]

The bottom line on biofuels' environmental effects

Rapid, large-scale deforestation to clear lands for biofuel crops, supposedly being done to reduce CO_2 releases into the atmosphere and benefit the environment, will actually increase CO_2 releases in the next years and decades, contributing to global warming rather than working against it, and increasing the risk of extinctions rather than reducing them. Even where biofuels might give a net energy yield, the environmental costs are too high to justify the production of these fuels, even if the monetary cost were low, which it is not.

Biofuels' effects on the pocketbook, direct and indirect

We previously talked about farmers feeding their animals trail mix because feedstock crops have been crowded out by agrifuel crops, making feedstock expensive and hard to get. The same thing is true for our own food crops, with scarcity driving up prices. We feel these higher costs directly.

But it is costing us even more than we realize. Biofuels are expensive to produce. The former head of the Malaysian Palm Oil Association, M. R. Chandran, was quoted in the *Wall Street Journal* in 2007 as saying that crude oil would now have to be as much as $130 a barrel before palm-oil-based biodiesel is competitive.[52] Biofuels are therefore heavily subsidized—by our tax dollars.

So why are biofuels so popular?

Subsidies make biofuels seem less costly than they are. They therefore seem attractive, reasonable, and cost-effective. Precisely how heavily they are subsidized is hard to say, because there are so many ways that governments provide financial support for agriculture in general and for biofuels in particular, some direct, some indirect and hard to uncover.

Big ag—corporate agriculture—is accustomed to big handouts and knows how to get them. As an example, a study done by researchers at the University of California shows that federal subsidies for rice totaled $269.5 million in 2002, exceeding revenue to the farmers from direct sales, which were $214.9 million that year. The study concluded that "rice growing in California could not be done without federal subsidies. This is basically a handout program."[53]

Subsidies can make crops for fuel more profitable than crops for food. A May 2007 article in the *International Herald Tribune* stated that "bioenergy crops have now replaced food as the most profitable crop in a number of European countries. In [Ardea, Italy], for example, the government guarantees the purchase of biofuel crops at £22 per 100 kilograms, or $13.42 per 100 pounds—nearly twice the ... rate on the open market last year. Better still, European farmers are allowed to plant biofuel crops on 'set-aside' fields, land that EU agriculture policy would otherwise require them to leave fallow to prevent an oversupply of food."[54]

In China, where there is a major commitment to developing biofuels, estimates are that the processing of 100% bioethanol costs about $150 per ton, and this does not include costs for purchasing feedstock. Subsidies are high in China—somewhere between $200 and $300 a ton.

If biofuels did not get big subsidies, in most countries few could afford them. In the United States, corn-based ethanol gets $3 billion in federal and state subsidies each year. Without these, corn-based ethanol production would either greatly decrease or stop altogether. Direct subsidies are $1.00 a gallon for biodiesel produced as an agricultural crop directly for use as fuel, and $0.50 a gallon for biodiesel produced otherwise (such as from wastes).[55] And these are just the *direct* subsidies for biofuels. In addition, their production benefits from the elaborate and complex subsidies for agriculture and from a variety of indirect subsidies that are often difficult to learn about and add to the financial balance. For example, in California in the 1980s, the city of Los Angeles was paying about $300 an acre-foot for water purchased from the Colorado River system, but farmers in the state's great central valley were getting that same water for $1 or $2 per acre-foot. (An acre-foot, the amount of water to cover an acre of land 1 foot deep, is a standard U.S. measure of water quantity.)

Senator John McCain reported about these subsidies, concluding that the cost to produce a corn-ethanol energy equivalent for gasoline was $4.77 per gallon, while the cost to produce a liter of gasoline from fossil fuels was only $1.27 per gallon.[56]

Federal and state subsidies for ethanol production go mainly to big corporations, not to individually owned farms.[57] One of the most careful analyses of agrifuels, by David Pimentel and Tad W. Patzek of Cornell University, determined that "several corporations, such as Archer Daniels Midland, are making huge profits from ethanol production."[58] Furthermore, they estimate that if you include the increased cost of food, "the costs to the consumer are greater than the $8.4 billion/yr used to subsidize ethanol and corn production." The National Center for Policy Analysis estimated that ethanol production added more than $1 billion to the cost of beef production.[59, 60]

In sum, subsidies explain the popularity, and profitability, of biofuels. Without government subsidies, biofuels would not be profitable in the United States and many other nations—the cost efficiency would be too low. And in the United States, the primary beneficiaries of biofuel subsidies are large corporations, not individual farmers.

Could great advances be made in biofuels' energy production?

To answer this, we have to go back to a point made earlier: that photosynthesis is a natural and ancient process "invented," so to speak, by bacteria and then "borrowed" and "improved" by algae and green plants. It is a way of using solar energy.

The first research project I ever did in ecology dealt with just how efficient natural vegetation could be as energy factories. I studied how much energy the natural vegetation in an old field, abandoned by a farmer only a year earlier, was able to convert to its organic matter compared with the amount of sunlight available. According to ecological theory of the time, natural ecosystems were most productive in their earliest stages, and therefore this abandoned old field should have been tops in storing energy. In fact, however, this mixture of weedy plants stored just a bit more than 3% of the solar energy that fell on the field during the entire year.[61] That's an indication of what can be expected from natural vegetation.

The old saying that it's hard to improve on Mother Nature is going to be true for biofuels. As amazing as photosynthesis is, it has its limits, and it is unlikely that the net yield from even intense agriculture is going to do better than that old field. Why? For starters, photosynthesis can make use of only a limited percentage of sunlight—enough to cause the photochemical reaction that leads to energy being stored in organic compounds, but not so much that the energy can seriously damage and destroy living cells.

So think about biofuels as a weaker form of solar energy. Its efficiency is going to be not much better than 3% of the sunlight and certainly less than 10%. Contrast that with what we discussed in the chapters on wind and solar. For example, today's most efficient photovoltaics "fix" 20% of the sunlight—one-fifth of all sunlight that falls on them goes out as usable electricity.

Should biofuels be wholly disregarded?

Despite all the negatives, and because there is plenty of pressure from nations lacking petroleum reserves and from corporations seeking profits from agriculture, it's a good idea to keep natural vegetation in mind as one possible sustainable approach to biofuels. A study of a natural prairie in Minnesota indicates that fuel made from a mixture of wild plants could yield "51 percent more energy per acre than ethanol from corn grown on fertile land....This is because perennial prairie plants require little energy to grow and because all parts of the plant above ground are usable."[62]

Biofuels attract a lot of government funding and private investment in countries such as China and Brazil that lack large petroleum reserves and have a lot of land that is either not in use for food crops or only marginally productive for food. For these nations, biofuels offer at least a short-term gain, although at a large long-term cost in terms of food production and the environment. The Chinese Academy of Forestry has identified 1,553 species of oil-bearing woody plants that might be used to make biodiesel fuel.[63] Companies in the Hebei province of northern China are said to be processing more than 20,000 tons per year of pistachios for biofuels. The claim is that the nuts have an oil content greater than 40%, which makes them a good source, and that pistachio trees can be grown in mountains and hills unsuitable for farming. The company says there are 160,000 acres of such land available and that the pistachio yield averages slightly more than seven tons per acre per year.[64]

In the U.S. today, hundreds of biodiesel plants produce fuel in many states, and among owners of small farms there is a rapidly growing interest in biofuels, especially biodiesel fuels. Cooperatives are springing up around the country, with members who are interested in sustainable agriculture and improving the environment, and who view the benefits of local production of biofuels, especially biodiesel, as outweighing doubts. One such organization, the Piedmont Biofuels Cooperative of North Carolina, writes on its website: "Our mission is to lead the grass-roots sustainability movement in North Carolina by using and encouraging the use of clean, renewable biofuels." [65]

College courses in biofuels are springing up as well, such as the Biofuels Program of Central Carolina Community College.[66]

This is an interesting situation, in which one cannot but applaud the enthusiasm and desire to do good locally. The question is whether, even locally, with transportation minimized and with many related tasks done sustainably, the net result can be energy-efficient, environmentally sustainable, and economically profitable.

The scientists who analyzed the production of ethanol from sugarcane in Brazil and corn in the United States, and concluded that the environmental and economic consequences outweighed any benefits, nevertheless suggest that "the ethanol option probably should not be wholly disregarded." They explain:

> The use of a fuel that emits lower levels of pollutants when burned can be important in regions or cities with critical pollution problems. Also, in agricultural situations where biomass residues would otherwise be burned to prepare for the next

planting cycle, there would be some advantage in using the residues for alcohol production. However, further research should be done to improve the conversion process. Considering that, eventually, petroleum may no longer be available in the amounts currently consumed, one must conclude that substitution of alternatives to fossil fuel cannot be done using one option alone. It will prove more prudent to have numerous options (e.g., ethanol, fuel cells, solar energy), each participating with fractional contributions to the overall national and global need for fuel energy.[67]

The bottom line

- Fuels produced by algae and soil bacteria seem to hold the most promise for culturing and growing species with the goal of producing usable energy.

- Agrifuels—crops grown on the land with the goal of producing fuels—cannot fill a significant percentage of our energy needs. Currently, even the most optimistic estimates indicate that it would take all the cropland planted today in the United States just to fuel all of our automobiles.

- Even when the net energy balance is positive—that is, when we get more energy from a crop than we used to produce it—biofuels seem to produce more environmental damage than benefits.

- Burning waste as fuel is a valuable way to get rid of undesirable chemicals, but waste, too, can never satisfy more than a small percentage of our total energy requirements.

- If we seek to reduce the net release of carbon dioxide into the atmosphere, it appears better to slow deforestation than to grow crops for fuel. An acre of forest stores more carbon than would be gained by converting an acre of cropland from crops for food to crops for fuel.

- Turning biological waste into fuel is in general a good idea, even when energy efficiency is less than 1—it's just a smart way to get rid of waste. But you can't estimate the energy efficiency or carbon benefits as if the fuel just suddenly appeared at a gas pump with no costs associated with its original production.

- For household use where no other fuels are available, firewood, animal dung, and other local sources are necessities for heating and cooking. The problem is how to make these into sustainable products that minimize environmental damage.

- Crops grown for biofuels today are not cost-effective, are damaging to the environment, and threaten biodiversity. In the future, as more is learned about them, they will probably have a niche role and be important in certain nations.

- Burning organic wastes and using firewood that is harvested following sustainable practices are good things to do and will play a definite if not major role in world energy supplies. Firewood will remain important in rural areas and in poor nations, at least until small windmills and small solar installations take over.

Section III

Designing an energy system

Now that we have explored each of the sources of energy, we can turn to the important and more general question: How can we design an energy system for the future that will provide abundant energy, and much greater energy independence, with the best combination of energy sources and the greatest reliability and cost efficiency? This involves some larger-scale questions, including how we will transport energy, how we can use energy to transport ourselves and our goods most efficiently, and how we can improve the energy efficiency of our buildings. Finally, in the last chapter I discuss possible solutions and try to put the entire story together.

10

Transporting energy: the grid, hydrogen, batteries, and more

FIGURE 10.1 On June 10, 1999, a pipeline transporting gasoline exploded near Bellingham, Washington, a rare accident, killing three boys and causing considerable local environmental damage. Some 230,000 gallons of gasoline were spilled and one and a half miles of Whatcom Creek were damaged, killing an estimated 100,000 fish. Smoke from the explosion rose six miles. (© AP images)[1]

Key facts

- The U.S. energy transport network is huge. Some 90,000 miles of oil pipelines, 2 million miles of natural-gas pipelines, and 700,000 miles of electrical transmission lines transport much of the energy from where it is obtained to where it is used.

- America's natural-gas pipelines have had a good safety record, but all energy transportation systems are vulnerable to terrorism and accidents, which could have far-reaching effects.

- Ironically, the most technologically advanced form of our energy—electricity—has the most outdated, inadequate, and vulnerable transport network, the electric grid.

- If present trends continue, peak electricity demand will be unmet in five years for most of the United States unless we rapidly expand our transmission system. A hydrogen economy, where a society produces and uses energy primarily or largely in the form of hydrogen, is a popular proposal today. But the United States lags other nations—including Japan, Germany, and Denmark—in research and development to create such an economy.

Pipelines: one way to get energy where you need it

When we think about transporting energy—if we think about it at all—we picture high-tension power lines marching across the landscape. Aware of it or not, however, electricity is not the only form of energy that must be transported long distances before we use it. Gasoline, for example, travels a long way from refineries via pipelines, rail cars, and trucks before it gets into your car's tank.

One of the many problems associated with Americans' dependence on fossil fuels is that the places where the fuels originate are usually far removed from where they are most heavily used. For example, East Coast states, with their high human populations, receive 60% of the refined oil products shipped within the nation and almost all the refined oil products imported into the nation.[2]

Mostly it takes a disaster or a hugely inconvenient disruption to make us suddenly aware of energy transportation. Sometimes the attention-getter is a major oil spill, like that of the *Exxon Valdez*, or news of a spectacular explosion when a pipeline bursts or a railway car or truck carrying gasoline overturns, as happened on June 10, 1999, in Bellingham,

Washington. It was hard to miss this explosion if you were in Bellingham or nearby—the smoke rose six miles into the air, and more than a mile of gasoline several inches thick slid down Whatcom Creek in the town. The next year, in August, 2000, a gasoline pipeline exploded near Carlsbad, New Mexico, and killed 12 members of a family camped nearby. As a result, in March 2002, Senators John McCain and Patty Murray authored the Pipeline Safety Improvement Act as an amendment to the Senate energy bill (S 517).[3]

When I talk with people about alternative energy sources such as solar and wind, inevitably someone asks what we're going to do about running cars when all we are producing is electricity. Once in a while, someone may ask about the electrical grid, especially right after a major blackout, but I can't remember anyone ever asking questions about transporting natural gas, gasoline, diesel, jet fuel, or coal within the United States. If there's an oil spill somewhere in the ocean, or if off-shore drilling comes up in the news, then people talk about local effects of a spill, but rarely about the national or international transport of oil.

There are two key points here. First, all forms of energy have to arrive at the place where we want to use them or can use them. (We can sometimes go to the source of the energy, as do farmers taking their grain to a medieval watermill.)

Second, energy can be converted into forms that are more easily transported, although the conversion always entails some loss. With the invention of the fuel cell, the conversion of electrical energy and chemical fuels became practical for many modern technological applications. An electric current passed through water separates H_2O into hydrogen and oxygen. Although hydrogen is highly explosive and therefore hard to package, it is one of the best fuels. It can be combined with carbon to make methane, the simplest hydrocarbon (one carbon atom combined with four hydrogen atoms). Add an oxygen atom to methane in the right way and you have ethanol, alcohol that can power your car. In this way, the energy from sunlight, first converted to electricity, is transferred as energy stored in a gas or liquid fuel.

The processes can also go the other, more familiar way—as many power plants do all the time, and as those convenient little home generators do: Use gasoline, diesel, oil, natural gas, or coal to run an electric generator, converting the energy stored in those gas and liquid fuels to AC or DC.

Each form of energy that we use to power our civilization has a transportation network. The networks are huge, and as the accompanying illustrations show, each network is surprisingly complex (Figure 10.2).

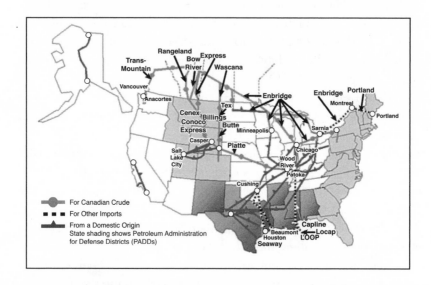

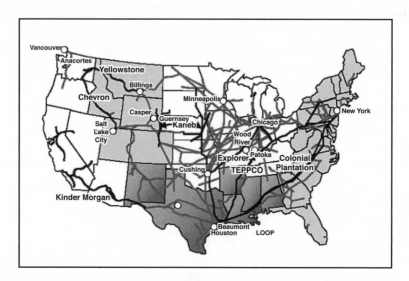

FIGURE 10.2 Oil pipelines in the United States: (Top) The big trunk lines.[4] (Bottom) The smaller refined-oil lines.[5] *(Allegro Energy Group)*

The U.S. petroleum pipeline

For petroleum alone, there are 55,000 miles of main "trunk" pipelines and 30,000–40,000 miles of smaller "gathering" pipelines in the United States, including both underground and aboveground pipes.[6] Petroleum accounts for about 17% of all freight moved in the United States, and the pipelines carry about two-thirds of all that petroleum.[7] It's hard to imagine how much petroleum is transported—it's another of those giant numbers that populate discussions of energy—but let's try to picture it. According to one analysis, it simply couldn't be done by truck or train. "Transport for high volume/long distance shipments are so daunting as to be impractical. Assuming each truck holds 200 barrels (8,400 gallons) and can travel 500 miles per day, it would take a fleet of 3,000 trucks, with one truck arriving and unloading every 2 minutes, to replace a 150,000-barrel per day, 1,000-mile pipeline."[8] And if all this were to go by rail, "Replacing the same 150,000-barrel per day pipeline with a unit train of 2,000-barrel tank cars would require a 75-car train to arrive and be unloaded every day, again returning to the source empty, along separate tracks, to be refilled."[9]

Transporting natural gas

Right now, 19% of electric power in the United States is produced by burning natural gas, and this is expected to increase to 23% by 2016. The natural gas used in the United States flows through 300,000 miles of major trunk lines and 1.9 million miles of smaller lines, including those that deliver gas to your house and to 69 million other users of this fuel (Figure 10.3). Some areas of the nation depend quite heavily on natural gas for electricity. Texas gets more than half of its electricity from natural-gas-powered plants. Florida, California, Arizona, and parts of the Northeast—areas with high populations—are also very dependent on natural gas for electricity. A disruption in the natural-gas transportation network therefore could affect both heating and electricity in a large part of the U.S.[10] The North American Electric Reliability Corporation agrees, saying that "disruptions in the supply or delivery of natural gas could have a significant impact on the availability of electricity" and that some measures to provide protection against such events are in development. These include more storage units as well as "alternate pipelines, expanded dual fuel capability, fuel-conservation dispatching, and increased coordination with gas pipeline operators."[11]

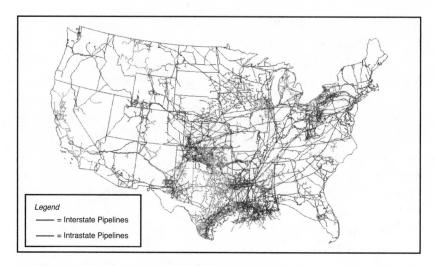

FIGURE 10.3 Natural gas pipelines.[12] (*DOE/Energy Information Administration/Office of Oil & gas, Natural Gas Division, Gas Transportation Information System*)

Remarkably, the natural-gas delivery network has been one of the safest forms of transportation of any kind, with only 12 fatalities in one year, 2002, during which there were 42,000 deaths on highways and a total of 2,000 deaths from aviation, boats and ships, and railroads.[13, 14] According to the American Gas Association, gas companies spend $7 billion a year to maintain these pipelines. Natural gas also travels in a liquefied state, which requires that the gas be highly compressed. This is the way it is also transported across oceans among nations, and it is much more controversial because of the risk of explosions and vulnerability to terrorism.

Advantages and disadvantages of the pipeline system

Like air travel, petroleum transportation has hubs and spokes. New York City is one of the major hubs for importing and transporting oil, as is otherwise little-known Cushing, Oklahoma, along with Chicago, Los Angeles, and several areas along the Louisiana-Texas coast. Oil spills are of particular concern for hubs with high resident populations.

It is important to note that not all the U.S. states are connected to each other by pipelines for either oil nor gas. California has no pipeline from other states, and New England has no pipeline connection to the rest of the nation—fuel arrives there by barge. This means that a large portion of the U.S. population lives where the least expensive and most efficient oil-delivery system isn't available.

The good news: Although the amount of material moved is huge, the cost per barrel or gallon is low. For example, in 2001 it cost only about 2.5¢ per gallon to send gasoline from Texas to New Jersey through pipelines. The cost is much higher by train, truck, and even by barge, as is the amount of energy expended to move the fuel by those means.[15] In case you were wondering, oil moves about 3 to 8 miles per hour in the pipelines, so it takes two to three weeks for oil to get from Houston to New York City. This lag might create supply problems in an emergency, an argument in favor of going the electricity route.

Transporting electricity: the grid, the smart grid, or no grid?

Going the electricity route—trying to obtain, transport, and use as much energy as possible in the form of electricity rather than using that energy to make gas, liquid, or solid fuels—has its own problems, as we saw in the book's opening story about the great 2003 blackout. Let's start with this simple fact: The world's largest machine is the U.S. grid system. Actually, it's three systems: an eastern grid that covers the eastern two-thirds of the United States and Canada; the Electric Reliability Council of Texas, which covers Texas; and a western grid that takes care of the rest of the United States and Canada that has grid connections.

In the U.S., the grid was originally developed only as an emergency fallback. It was built mostly in the 1930s through the 1950s to provide emergency power as needed and extend electricity to the smallest farms in the most rural areas. Today, the grid includes more than 700,000 miles of transmission lines[16] and 250,000 substations and has become the primary way of transporting electrical energy. With about 60% of its equipment more than 25 years old, one of the few things on which most energy experts agree is that the electrical grid is badly outdated, likely to fail, and in need of both major repairs and technological updating. (This is true even taking into account that electrical machines tend to be longer-lived than many others, such as internal combustion engines.) Bill Richardson, Secretary of Energy in the Clinton administration, said the U.S. grid was "third-world."[17]

How does something go wrong with the grid? When the electrical energy transmitted exceeds the amount that the wires can carry, they overheat and may sag or break, and if this doesn't happen, the transformers and other devices blow out. As the 2003 blackout illustrated, when one part of the system goes down, another becomes overloaded and fails, until the entire system crashes like a bunch of dominos. Also, the grid transmits alternating current, not direct current, and this must be generated precisely at the standard 60 cycles a second. If not, the entire system can get out of phase, overheating can occur, and the grid will fail even if the total amount of electricity flowing over the wires has not surpassed the maximum. The grid's electrical load has grown even greater since the blackout of 2003.[18]

The evidence provided so far in this book favors solar and wind energy, but this requires us to improve our ability to transmit electricity long distances over electric power grids and to put more emphasis on off-the-grid applications and what are known as *microgrids*, which transmit the electricity locally to a number of users over transmission lines. An analysis of the problems with the present electrical grid by Roger Anderson and Albert Boulanger of the Lamont-Doherty Earth Observatory, N.Y., published in *Mechanical Engineering Power & Energy*, concludes that "the present U.S. electric grid will not work on any scale—local, state, national, or international—at the higher loads and more diverse generation sources required in the future, let alone if the terrorist threat becomes more severe. Failing to upgrade the system will leave us unprepared and, ultimately, in the dark."

But little is happening. According to the nonprofit North American Electric Reliability Corporation, which assesses such things, only 2,000 miles of electrical transmission lines were added in 2006, less than 1% of the total and much less than what is needed, not even considering the need for repairing, replacing, and upgrading existing lines.[19]

One of the most important uses of fossil fuels in the next years will be to provide power that can be brought online at times when demand exceeds the supply of electricity generated by wind and solar. *Microturbines*—basically, the same engines that power jet aircraft—are right now used for this because the engines can be brought up to speed quickly to provide electricity when there is a sudden increase in demand.

A smart grid

Experts also favor a *smart grid*. Several advocacy groups have emerged calling for this, including the Galvin Electricity Initiative and the GridWise Alliance. The former is the brainchild of Bob Galvin, retired CEO of Motorola Corporation, who has made this a major activity since leaving that company. He argues that power outages cost the United States $150 billion a year and that the smart grid would prevent them with the use of automatic switching systems that involve computer and Internet-like communication and control. The smart grid will involve advanced management of electrical devices such as home water heaters, whose energy use could be automatically reduced when overall grid electrical demand surged. It would also have the capability to turn the charging of electrical vehicles on and off so that charging takes place primarily during off-peak times.[20]

Think of the transition from *dumb* grid to *smart* grid as similar to the transition from telephone lines in the first half of the 20th century, with operators handling all the calls, to the present cell phone world of phoning, text-messaging, game-playing, GPS, and the ability to contact your computer from anywhere.

The smart grid moved out of the idea stage in 2008 when Xcel Energy Corporation began installing a test system in Boulder, Colorado, that included *smart meters* that tell a customer how much electricity he uses and makes possible real-time adjustments. The idea is that this information can motivate people to reduce their use of electricity.

Surprisingly, redundancy has not been the common approach for the grid. As a result, using today's grid is a little like flying big commercial airplanes on one engine all the time, with your fingers crossed, just hoping that nothing goes wrong. Actually, substations are quite vulnerable, not only to lightning in a thunderstorm, but even to a squirrel that has found its way into a dangerous place and steps across two high-voltage wires.

A little-discussed danger is the possibility of a terrorist attack on any of the energy-distribution systems, which are quite vulnerable. Ironically, the smart grid, with its Internet-like computer controls, might be even more vulnerable to cyberterrorism. Safeguarding their large-scale energy-distribution systems presents a major challenge to the United States and other developed nations. This is all the more important because, as emphasized throughout this book, an adequate energy supply is fundamental to modern technological societies, any of which could be crippled at least temporarily by major disruptions in energy distribution. Although advocates for improving the grid discuss this, it remains one of the least publicized of the major issues about energy supply.

One of the solutions to disruptions of an electrical grid, including terrorism, is to train the grid operators much as airline pilots are trained, using sophisticated computer simulations so that they can experience and learn how to deal with rapid surges in demand. You may recall that one of the major causes of the widespread system failure in 2003 was that grid operators found themselves unable to respond quickly enough and get power companies to cooperate.

Advocates of the smart grid call for a large-scale integrated system of energy production, transmission, and storage, including novel kinds of energy storage, such as huge flywheels and underground compressed air in caverns and superbatteries and elevated water reservoirs.[21] They also call for novel, experimental methods of energy transmission, such as low-temperature superconductors. The first experiment with this kind of transmission took place in 2008 at Brookhaven, Long Island, New York, where the $60 million Holbrook Superconduction Project started 138,000 volts of electricity flowing along a half-mile of wires that were cooled to minus 371° centigrade by liquid nitrogen and were no bigger in diameter than spaghetti.[22]

The U.S. Department of Homeland Security reached an agreement with Consolidated Edison Corporation in 2007 to install superconducting cables beneath New York City to connect two Manhattan substations (big transformers that change the voltage of electrical currents) so that if one burns out, the other can take over. The superconducting cable for Manhattan is in the planning stage and is supposed to be installed and running in December 2010, but this is not certain. One reason that superconducting cables are planned is that so many underground wires, cables, and pipes exist in Manhattan, let alone subways and train tracks, that little room remains for massive new cable systems. A second reason is that heat generated by standard transmission lines would create problems in the crowded underground.

How much will it cost to repair, restore, and develop the electrical grid? According to the Edison Electric Institute, it will cost at least $450 billion (Figure 10.4). But this does not include all the costs of the smart grid or more exotic developments like superconducting cables.

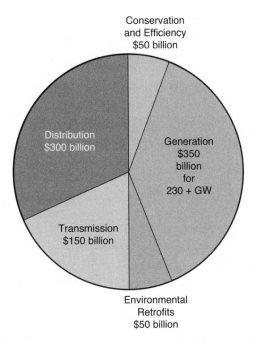

FIGURE 10.4 Costs to improve the electrical grid *(Edison Electric Institute estimates).*

No grid?

Our use of energy in the future will involve a greater degree of independence from the grid and from major national energy networks. This will be made possible by solar and wind energy and by the development of local microgrids, referred to earlier. Our future energy supply will also involve the integration of different forms of energy and energy transportation—including, in particular, the conversion of electrical energy to hydrogen. Some call for a hydrogen economy, meaning that hydrogen would become the fuel of choice and, along with electricity, the primary means of transporting energy. The National Renewable Energy Laboratory (NREL) claims that the United States could convert to a hydrogen economy in ten years.[23]

This would seem an extreme challenge, with the nation not ready for it. But Iceland and Japan illustrate the potential for moves in this direction. Iceland has several filling stations that provide hydrogen as a fuel for cars. In Great Britain a plan is in development to make the Shetland Islands independent of fossil fuels by producing electricity and hydrogen from wind.[24] Denmark built Europe's first wind-to-hydrogen facility on the island of Lolland.

In sum, however, the idea of a hydrogen economy is controversial and largely untested. Although there is much talk and informal journalism about a hydrogen economy, and the idea has been the topic of several popular books, including Jeremy Rifkin's *The Hydrogen Economy*,[25] the present reality is that such energy systems are only experimental and small-scale, and little is happening in the United States.

The bottom line

- The transport of energy is one of the keys to energy independence, security, and our standard of living and way of life. But compared to questions about whether to go nuclear, switch to biofuels, or keep searching for more fossil fuels, it gets little attention.

- The future of energy in a technologically sophisticated nation will involve better integration of energy networks, a smart electrical grid, and greater use of microgrids—producing and using energy locally, within a relatively small area.

- Too little research, development, and imagination are focused on transporting energy. It should be one of the major areas of innovative energy research, but it is not.

- The move away from fossil fuels, no matter what kind of energy becomes primary, will lead to increased production of electricity. Conversion of electrical energy to gas and liquid fuels will be necessary, but methods and installations are presently woefully inadequate.

11

Transporting things

FIGURE 11.1 General Motors Hummers weren't exactly flying off the lot in 2008. That year, movie and television stars discovered electric and hybrid vehicles, and Hummers were left sitting largely unvisited on new-car lots. *(Photo by Daniel B. Botkin)*

Key facts

- Almost one-third—28%—of the energy we use in the United States is for transporting ourselves and our goods.

- Because our transportation choices are flexible, transportation is a key way to start saving energy quickly.

- In the United States, trains move 40% of all freight. But this generates only 10% of the total U.S. freight revenue because it is so much cheaper and more energy-efficient to ship things by train than by truck.

- Cars and light trucks together use 63% of all energy used for transportation; trains and buses together use just 3%.

- More than 6% of all energy used in the United States is simply to transport coal, mostly so it can be burned as fuel but also as an ingredient of steel.

- Air freight accounts for less than 1% of the total freight moved in the U.S., but 12% of the total U.S. transportation revenue.

- The most fuel-efficient way to move people around on land today is by intercity bus. Trains are next. But for transporting goods of all kinds, you can't beat ships for fuel efficiency.

The new status symbols: hard-to-get energy-saver cars

The big transportation news of 2008 was the rapid rise in gasoline prices, the subsequent flight from gas guzzlers, and the sudden popularity of smaller electric, hybrid, and hydrogen cars even among America's glitterati. Joely Fisher, star of Fox's *Til Death* TV show, managed to buy one of only 20 available BMW hydrogen cars. Brad Pitt and Angelina Jolie, Cameron Diaz, and opera star Placido Domingo also managed to get theirs. Meanwhile, basketball great Magic Johnson, *The Tonight Show's* Jay Leno, and America Ferrera, star of *Ugly Betty*, were driving the $1-million Chevy Equinox powered by fuel cells.[1] Rumor was that George Clooney, Jay Leno, Matt Damon, Brad Pitt, and Arnold Schwarzenegger were on the waiting list for the superfast-accelerating Testar all-electric car (0 to 60 in four seconds, range of 250 miles, top speed 130 mph).[2]

How could U.S. automakers not have seen it coming?

The mystery in all of this is why the big three auto companies of the United States didn't see this coming and didn't plan for it. General Motors hadn't had a full year of profitability since the early 1990s but kept manufacturing Hummers, big pickups, and SUVs long after all the rest of us had realized that almost all the cars around us on the highway were smaller, fuel-efficient imports. Between May 2007 and May 2008, sales of SUVs fell 38%.[3] Thus the real, fundamental transportation news of 2008 was the lack of foresight and planning, both by the big automobile corporations and apparently by much of the federal government.

It wasn't an obscure forecast that oil prices were going to rise rapidly about this time. As discussed in Chapter 1, "Oil," petroleum experts have been pointing out for years that when the time of peak oil discovery is reached and passed—estimated to occur between 2020 and 2050—the price of oil will go up. Thus warned, we didn't have to wait until we were actually in the midst of a fuel crisis, with demand significantly exceeding supply. The rapid rise in the standard of living and the economies of India and China were no secret either. Clearly, oil and gasoline weren't going to be cheap in the United States much longer.

A corporation that was thinking ahead would have tried to be ahead of the curve by focusing on technological development. Suppose Steve Jobs of Apple Computer, instead of inventing the iPod and the iPhone, and continuing to develop computers in advance of what the public was buying, kept trying to sell bigger and more expensive 1990s desktops forever. That seems to have been the Big Three automakers' approach when Toyotas, Hondas, and other foreign cars were zooming past them with ever more reliable and fuel-efficient models that soon pushed American cars out of their long-held first-place spot and left them in the dust.

Now we've got some catching up to do

Transportation uses 28% of all the energy used in the United States. It is a large percentage, but one that is readily changed. We have great flexibility when it comes to transportation, and it is therefore a key way for us to save energy quickly. When fuel was plentiful and cheap, we became careless and wasteful. Now we need to make wiser choices in how we use transportation energy.

Transportation basics: how? how much? and how efficiently?

According to the U.S. Department of Transportation (DOT), the use of cars and light trucks for personal transportation consumes 63% of the total energy used in the U.S. for transportation. (Cars use 35%; light trucks use 28%.) This is remarkable—our personal transportation in those fashionable pickups and in cars often carrying only the driver uses *almost two-thirds* of all energy used for transportation in the U.S. Other trucks, including the big semis, use just 16%, aircraft 9%, watercraft 5%, construction and agriculture 4%, pipelines 3%, and trains and buses together another 3%.[4, 5, 6]

Railroads can carry a ton of cargo 404 miles on just one gallon of fuel. (That's how the cost of freight transportation is measured, by the cost to move a ton a mile.) A train can carry a ton 10 miles for 1 kilowatt-hour. Remember, that's just the amount of energy needed to light ten 100-watt bulbs for an hour, and it's likely that in moving that ton a mile, a railroad at the same time also has at least that many 100-watt bulbs burning an hour. In the United States, trains move 40% of all freight, and they do it with such energy efficiency that rail freight charges are much cheaper than for trucks. As a result, the total revenue received by all U.S. railroads is only 10% of the total paid for transportation—$54 billion.

For air freight, it's just the other way around—planes move only a little of the freight, less than 1% of the total ton-miles, but at high cost, amounting to 12% of total transportation revenues.

As for boats, since their invention thousands of years ago, it has always been true that transporting freight by water is the cheapest and most fuel-efficient way to move it.

Coal makes up 44% of tonnage transported in the United States and 22% of the ton-miles, so transporting coal consumes 6.21% of all energy used in the U.S. *If we did not burn coal, and therefore did not have to transport it, U.S. energy use for transportation would decrease by 6.21%, or 1,821 billion kWh. Thus, moving away from coal is a double savings: It reduces the use of the dirtiest fossil fuel and increases the nation's energy efficiency.*

Improving the energy efficiency of transportation

It doesn't take rocket science to figure out some obvious ways to reduce the amount of energy we use for transportation. We can start with automobiles. Americans drive 3 trillion miles a year—10,000 miles a person for every man, woman, and child in the nation![7, 8] In doing so, Americans use 123 trillion gallons of fuel per year, or 412 gallons a person. If there is no change in miles traveled and average miles per gallon, then in 2050 the United States will use 173 trillion gallons of gasoline.

According to the U.S. Department of Transportation, the average miles per gallon (mpg) for new automobiles in the United States reached 24.3 in 2004 (the most recent figures available at the time of this writing), up just 7% from 1980. This was an improvement of only about one-quarter of a percent per year, way under what the automobile industry is capable of.[9] The most recent energy bill, the Energy Independence and Security Act of 2007, requires that by 2020 the average will be 35 mpg, a 44% improvement overall but still asking for less than 1% improvement in gas mileage per year.[10] If 35 mpg is still the average in 2050, when there will be an estimated 120 million more Americans, and if there has been no decrease in miles driven per person, then America will be at breakeven with today—burning about the same amount of gasoline each year. On the other hand, if cars got an average of 50 miles per gallon and the mileage per person dropped 50%, the amount of fuel used would drop to 42 billion gallons—34% of the amount used in 2007 and only 100 gallons per capita. *This would amount to a 10% decrease in the total energy used for all purposes in the United States in 2007.*

We saw a decline in miles driven in the late spring and early summer of 2008, in response to $4.00-a-gallon gasoline. According to the former U.S. Secretary of Transportation, Mary E. Peters, in the first quarter of 2007 Americans drove 20 billion miles less than in the first quarter of 2006, which would translate into a decrease of 2.7% in just one year.[11] Although there will be variations in the retail price of gasoline, we can expect the average price to keep going up unless government policies and subsidies force it down. Further hikes in the price of fuel will likely lead drivers to either drive less or trade in even more of the big gas-eaters for vehicles that are more fuel-efficient, as happened in mid-2008.

Thus, even without any improvement in mass transportation, even without any change in laws, a 50% decline in per-capita miles driven is possible by 2050.

Of course, this may not happen—people may adjust to higher gasoline prices, refigure their household budgets accordingly, and resume their old driving habits. Then, too, there's probably some minimum that could be reached in terms of miles driven per person each year unless mass transit expands and improves rapidly, with more and better trolleys, subways, and buses, and intercity high-speed rail and urban rail. Also needed are more and better bike lanes within urban areas and bike paths between urban areas. That's where the ways to reduce energy used in transportation become complicated, requiring imagination, innovation, and insights more at the level of rocket science.

Railroads are a big part of the solution

In the United States, 23,000 miles of intercity railroads provide access to 500 cities with 260 trains a day.[12] Unfortunately, in recent years too few people have traveled by rail, complaining that it is expensive and unreliable. At the time of this writing, however, rail travel was up considerably and the question was whether Amtrak could keep up with the demand.

When I started to write this chapter, I assumed it would be very expensive to upgrade existing railroads and build new ones, with their nicely graveled and cemented roadbeds and all those signals and switches and grade crossings. But to my surprise, when I talked with some experts on rail travel—including Tom Payne, director of the Ferroequus Railway Company Limited and founder of RailLink Ltd., Canada's third-largest railway—I found that restoring and building railroads isn't all that expensive. Let's go through the numbers.

Based on current costs, plans, and proposals, new high-speed railroad lines could cost as much as $2.5 million a mile to construct, but it could be lower in some situations. (That's for one track and all the signals, bedding, and other accoutrements.)[13] Add the cost to purchase the land or rights-of-way and the cost escalates to $20–40 million per mile, according to U.S. government information. Just restoring an existing rail line will cost from about $500,000 to $1 million per mile,[14] but where there's a need for a tunnel or bridge, and where the terrain has lots of ups and downs, the costs go up. These numbers give us a basis for making estimates, and taken alone they sound like a lot. If you don't like railroads or are in a business that competes with railroads (like making or selling automobiles), these are the kinds of numbers to throw around, without mentioning the costs of other transportation.

But when people begin to depend more heavily on rail travel, and upgrading and expanding existing rail lines and building new ones become necessary to meet increased demand, then it will become clear that the total construction costs are quite reasonable compared with the cost to improve other kinds of transportation. Based on estimates by the Feds and by railroad experts, the cost to build a new high-speed railway from scratch between Los Angeles and San Francisco or Sacramento, construction and equipment only, would be about $700 million. (This is the cost of building a route directly down the Great Central Valley, much better terrain for a railroad than the scenic route along the coast that the present railway follows.) Adding in the cost to purchase land would bring the price to $7–14 billion.

That may sound like a lot, but just building the Denver airport that replaced Stapleton in the mid-1990s cost $4.8 billion,[15] and $7–14 billion is actually a fraction of the estimated total cost of restoring the rest of America's infrastructure. The American Society of Civil Engineers' "Report Card on America's Infrastructure" estimates that $1.6 trillion will be needed over just a five-year period to get things back into shape, including bridges, tunnels, highways, airports, sewage lines, dams, hazardous wastes, schools, and navigable waterways.[16] A piddling railway between two of America's greatest cities, L.A. and San Francisco, would use up about a half a percent to not quite 1% including land purchases.

An energy-efficient United States will involve not only restoring and expanding passenger rail service but also keeping up with increases in freight transportation as the population and economy expand. The Department of Transportation estimates that freight traffic will increase 50% by 2020 and that meeting this increased need will cost $175–200 billion over the next 20 years.[17]

What would happen if the U.S. rail system essentially collapsed? According to the American Association of State Highway and Transportation Officials (AASHTO), if all freight moved by truck and none by rail, it would cost "an additional $69 billion annually," costs that would be transferred to us, the consumers. Moreover, the increased truck traffic would pound away at existing highways, which AASHTO estimates would require "an additional $64 billion in highway funds over the next 20 years" to maintain. Not to mention the need for new highways to handle all that extra truck traffic.

The government funds highways and air travel, with little if any for railroads

Although increasing rail traffic is a sure way to decrease energy use, government funding works to the opposite effect: In the mid-1990s, the federal budget provided more than $40 billion for the Department of Transportation to spend annually in highway and transit grants to states. The 2005 "Safe, Accountable, Flexible, and Efficient Transportation Equity Act: A Legacy for Users (SAFETEA-LU)," signed into law that year, provided $286 billion from 2004 to 2009 for highways, highway safety, and public transportation. The $286 billion included $815 million for the National Highway Traffic Safety Administration; $5.1 billion for highway safety programs; $517 million for the Federal Motor Carrier Safety Administration to provide grants to states for truck and bus regulation and enforcement; $100 million for up to five states to test various new ways to charge for the use of highways; $7.8 billion for "programs that address highway congestion"; $23.7 billion "to maintain and improve the National Highway System and to replace, rehabilitate, and preserve bridge and other infrastructure [of highways]."

What about railroads? The SAFETEA-LU did include $7.3 billion for "investments in buses, rail cars, and maintenance facilities," but only $900 million for Amtrak. At the time of this writing, this bill has not been reauthorized.

The Obama administration announced on February 1, 2010, that the core highway program would be funded at $42 billion per year in 2011—about the same as in the previous administration's 2010 budget but below the $67 billion actually spent by the federal government on highways in 2009—and that the Department of Transportation would spend $8 billion to develop the nation's first high-speed intercity transportation. But the Obama budget lowers funding for conventional rail from $11.1 billion actually funded in 2009 to a proposed $2.9 billion for 2011. This means that total funding for new high-speed rail and conventional rail is about $10.9 billion, an actual decrease from the $12 billion spent by the federal government in 2009.[18]

This set of federal policies, favoring highways and complaining about railroads, is not a free-market approach, although, ironically, it was the policy advocated by the George W. Bush administration, which claimed to believe in a free market. Under this continuing policy, highways will get much greater federal subsidies than railroads.

Similar policies have prevailed within states. For example, in his 2006 State of the State address, California governor Arnold Schwarzenegger proposed spending $222 billion for infrastructure development, which included $107 billion for transportation and air-quality programs—1,300 miles of new highway lanes and thousands of miles of bike and pedestrian paths, but just 600 miles of additional commuter rail, and no mention at all of high-speed rail. This for the nation's richest and most populous state.

We need more railroads and fewer cars and trucks

If the $40 billion given annually by the U.S. Department of Transportation to fund highways went instead to railroads, it could be used to build 16,000 miles of railway track, or 8,000 two-way rail lines, enough to cross the U.S. and more. Even if we assume that every inch of the land and rights-of-way for these rail lines would have to be purchased, that none of the new track would be laid on what is already government land—which would not be true—this would fund 1,000–2,000 miles a year.

Think about what this would get us. A new rail line could be built from New York City to San Francisco (2,900 miles), another from Boston to Miami (1,500 miles), a West Coast line from San Diego to Seattle (1,260 miles), from Seattle to Minneapolis (a northern route of 1,650 miles)—a total of 7,300 miles with just one year's allocation of what has been going to highways. And in the second year, we could add a Midwest north-south route from El Paso to Fargo (1,450 miles).

Alternatively, we could divide up the funds, half for intercity railroad and half for intracity and suburbs-to-cities rail transportation. According to railroad expert Tom Payne, light rail and regular railroads can be built for the same amount, although light rail is often more costly, he says, because of options chosen that aren't necessary for the basic transportation. (Light rail refers to intercity trains that are designed to carry lighter loads at lower speeds, but are more than trolley cars, like the trains in downtown Denver, Colorado.)

Like railways, air travel is vital to our economy

It has become fashionable among some environmentalists and "eco-friendly" advocates to focus on the use of fossil fuel by commercial air-lines and private jets, but it should be clear by now that their focus should be on automobiles and light trucks. Although aircraft use just 9% of the transportation energy in the United States, air travel has become essential to America's economy, and cities that lose commercial air serv-ice suffer economically. It is thus false economy and of little benefit to energy conservation to focus on reducing the use of fossil fuels for air travel.

As noted throughout this book, the solution to our energy problem involves a mixture of energy sources. Fossil fuels have been wonderful sources of energy, and what is left of them should be saved for the appli-cations in which they are most useful—for making plastics and other organic compounds, and for situations where oil and gas are especially well suited to provide energy. Until the time comes when jet fuel can be obtained in large quantities by using the energy from electricity to pro-duce small hydrocarbons, one of the most useful applications for fossil fuels will continue to be air travel.

Can we ever get ourselves out of our cars?

Getting out of our cars would save money and energy. Much more money is spent transporting people than freight in the U.S.—$1.01 tril-lion to move us around versus $580 billion to move our stuff in 2001, the most recent year for which there are data—a total of $1.59 trillion dollars spent on transportation.[19] In the U.S., highways account for 89% of the total passenger-miles, air travel accounts for 11%, and local transit (city bus and trains, etc.) just 1%.[20]

If things got so tough that people could travel in only the most fuel-efficient way, guess what that way would be. Surprisingly, right now the most energy-efficient passenger transportation, in terms of direct energy used to move one passenger one mile, is intercity bus. The Acela train is the next most efficient.[21] According to the Minnesota Regional Railroad Association, "On average, railroads are three or more times more fuel efficient than trucks."[22] Cars and trucks take three times as much energy per passenger-mile as the Acela, and air travel takes 4.48 times as much.

For every passenger who switches from traveling by car to the Acela, energy used per mile drops two-thirds. For every passenger who decides to travel on a standard U.S. train instead of by car, one-third of the energy per mile is saved.

Because it makes economic and energy sense to depend less on automobiles, trying to get Americans out of their cars has been a continuing effort by many experts and a primary emphasis of a number of university departments, such as George Mason University's School of Public Policy under the direction of Kingsley Hayes. Most of the efforts haven't succeeded very well, but a few have. I have some personal experience about this, which I would like to share with you.

First of all, I see it as a practical issue, not a moral issue. A growing debate about automobiles has taken on aspects of a moral argument, with some people seeming to believe that owning a car is immoral, while on the other side some appear to believe that it is a moral right, not just a practical convenience, to drive a personal vehicle. I can't say that I fall on either side of such a debate, but I do confess that I enjoy driving a car. I grew up in a small town where every teenager couldn't wait to be 16 and get a driver's license. For us boys, it meant a chance to drive around neighboring towns and try to pick up girls, or at least impress them. In those days of carburetors and points, before fuel injection and electronic ignition, I used to tune up cars. Today, I list among my close friends Lee Talbot, one of our greatest conservationists and also a world champion formula Ford and vintage car racer. I've crewed for him now and again (never being much of a help, but having a good time).

At other times in my life, I've lived happily without a car. I've ridden a bicycle since I was a youngster, and when I was a graduate student at the University of Wisconsin my only transportation was a bicycle and my feet. I remember that period as the time in my life when I had the fewest financial cares ever.

It was already obvious in 2008 that the more expensive gasoline is, the faster people move away from their personal vehicles. A July 28, 2008, front-page headline in the *Wall Street Journal* testified: "Funds for Highways Plummet as Drivers Cut Gasoline Use." All things considered, however, my conclusion is that for people to use personal vehicles much less than they do now, mass transportation must become convenient, attractive, affordable—and fashionable. If George Clooney, Matt Damon,

Cameron Diaz, Placido Domingo, Brad Pitt, and Angelina Jolie started to take the train and, for local trips, used bicycles and their feet, that would become fashionable and people would start to imitate them.

Doesn't this mean that a government intent on reducing energy use should help to make mass transportation more attractive rather than subsidize oil and natural gas corporations with the oil-depletion allowance and other ways that oil, natural gas, and highway travel are subsidized by the government?

Bicycles in cities

Today, my wife and I have an apartment in New York City and another in Florida and happily drive the 1,200 miles between the two. When we are in Florida, we drive everywhere because there's no public transportation to speak of and almost nothing except the ocean is within walking distance. In New York, however, we rarely drive, choosing instead to walk almost everywhere within walking distance and use public transportation for the rest, because, although crowded and noisy, it's the fastest way to get around. I bicycle for exercise in both places, but wish the Big Apple were better set up for cyclists. It has already improved greatly in that regard. It is well on its way to finishing the Hudson River Park, with a bike path that follows the waterfront from Battery Park at the southern tip of Manhattan and connects several miles north to a route that crosses the George Washington Bridge. In fact, you can go several hundred miles by bike from the middle of New York City if you want to.

Both my children, now grown and married, bicycle all the time. My daughter Nancy and her husband Mike sold their last car years ago. Mike and my son, Jonathan, bicycle to work. Mike and Nancy do their grocery shopping by bicycle. And when they visit us in Manhattan, they put us to shame by going on incredibly long bicycle rides. The last one, a night ride, took them up Eighth Avenue, one of the busiest Midtown thoroughfares, and then over to the East River, where they enjoyed the spectacular nighttime views of the city while bicycling back and forth across ten different bridges.

I confine my bike riding in New York City to the separate, park-like bike paths because riding in the streets is too dangerous. I've seen car/bike accidents, and friends and relatives have been hurt bicycling in cities. I would love to bicycle more in the city—in fact, I prefer it to walking, driving, or taking a bus or subway, as long as it isn't raining or snowing. When I meet neighbors bringing their bicycles up the elevator in our apartment building, I try to strike up a conversation about where they've been and how they feel about bicycling in a big city. Almost everybody tells me that they realize it's dangerous, but they love it, so they do it anyway.

My conclusion is that the number one way to get more people out of their cars and onto bicycles is to make it safe to do so. The methods—well known, straightforward, and in use in many European cities—include bicycle lanes either separated from motor vehicle traffic by a cement curb or on a route entirely their own.

New York City recently took a tentative step in this direction by turning an entire lane of busy downtown Ninth Avenue into a bike lane, with cement curbs separating the bikes from motor vehicle traffic. I wish the lane were longer and that the entire city had them.

Of course, just as large numbers of motor vehicles on crowded city streets require driver adherence to strict traffic rules, so will increased numbers of bicycles. Right now, too many New York City cyclists seem oblivious to traffic rules, including stoplights, and routinely endanger pedestrians and themselves.

To be honest, I love to travel, and I haven't met a form of transportation I haven't liked. Trains, planes, cars, buses, ships, barges, canoes, kayaks, freight steamers in the Philippines, Amazon River boats—for me, again, it isn't a moral issue but a matter of convenience, practicality, and fun. But it's clear that we can't go on driving as we have. There just isn't enough cheap energy anymore, and also the heavy traffic negatively affects the quality of our lives. Riding down the Hudson River bike path on a Friday afternoon, I watch in dismay as thousands of cars, most with just one occupant, inch their way north at the end of the workweek, barely moving while their tailpipes spew exhaust that forms a haze over the city.

Cities that are especially friendly to pedestrians and cyclists

I lived in Portland, Oregon, as part of a project I once did, and loved the way that city made it pleasant to be a pedestrian. One way it did that was with pedestrian walkways cutting midway through some of the city blocks, paths planted with trees, shrubs, and flowers, and with miniparks along the way. This led me to imagine a city of the future where bicycle paths and walkways followed such routes. Of course, in cities like New York, where land is so valuable and so many buildings are already in place, this is going to be difficult. But it's not impossible. Highway engineers working with landscape architects could lead a major step forward.

Another key that is growing in popularity is to make it possible and convenient to use a bicycle for short trips even if you don't own one. Paris, France, has started to do this with a program that provides bicycles at many locations for a small fee and lets you pick up a bike at the location nearest to you and drop it off near where you're going. Amsterdam has also adopted this program, and it's highly popular.

Carless cities: what more can we do?

If offering attractive alternatives doesn't make a big enough dent in inner-city traffic congestion, what more can be done? Nobody seems to have worked it out yet. One approach is "congestion pricing"—fining those who use personal vehicles in the busiest parts of cities at the busiest time of the day or week. One example is the charge to drive in downtown London, an approach that New York City's Mayor Bloomberg tried to adopt for Manhattan, but the New York State legislature nixed it in 2008.

That's the "big-stick" approach, and it appeals to some because it seems simple to just fine people for driving. But kinder alternatives— "carrots" that work—have admittedly not been easy to find. In New York City at the time of this writing, the Metropolitan Transportation Authority had a big deficit, and badly needed improvements in the city's huge subway system weren't happening. New York City's subways work amazingly well in terms of transporting a lot of people very quickly, and this system is the world's largest and one of the few that operates 24 hours a day. However, they are screechingly, ear-splittingly noisy, often jam-packed, and none too clean. Also, many stations are far underground and have minimal people-movers to get crowds up to the street. If you want to tempt drivers out of their comfortable cars, it would help if this subway system could become as quiet and pleasant as the metros in Washington, DC, and Paris.

Many major cities are on seacoasts or major rivers—that's because water transportation was so important that cities were founded at good river and ocean junctions. Ferries used to be common, but they too have become unfashionable and mostly abandoned in the United States. Still, there's hope. A small comeback is happening in New York Harbor with yellow Water Taxis and other subsidized ferries taking people across the Hudson and East River much faster than they'd get there by bus and subway, or even by car sitting in long rush-hour lines at tollbooths (and then looking for and paying dearly for parking).

Which brings us to another approach: building or rebuilding cities so that cars just can't get into them—or with few if any parking spaces—and at the same time improving bicycle and pedestrian pathways and various kinds of public transportation. You will probably be surprised to learn that the biggest close-to-carless area in a major U.S. city is not in one of the ecofriendly cities on the West Coast, like Portland. It's New York City's Roosevelt Island, a two-mile-long, 147-acre island in the East River. The main transportation to and from Manhattan for the island's 10,000 or so residents is by subway or by aerial tram across the river.

Most other car-free areas of cities are historic districts like medieval portions of European cities, or newly built planned communities like Vauban, 1,700 houses on what was a military base in Freiburg, Germany; its 4,700 residents accept streets too narrow for most automobiles. In Great Britain, Prince Charles has promoted car-free parts of cities, but this has met with considerable criticism.

Ironically, as traffic jams decrease, people have less reason to abandon their cars. As long as there are highways, broad avenues, and freeways/interstates/turnpikes, the traffic level will tend to have a negative feedback and the amount of traffic on them will tend to stabilize—the lower the traffic, the greater the growth in traffic; the worse the traffic jams, the greater the decline in traffic. Thus, part of the solution has to be to stop building ever-broader streets, avenues, and freeways, and instead put transportation money into light and heavy rail, as well as bicycle paths and park-like walkways.

A further note: microgrids can help

The increase in microgrids, described in the Chapter 10 on transporting energy, would also lead to an overall decrease in the transportation of freight and people, because energy would be produced and used locally, and people would live nearer to their jobs. How much of an energy savings this might yield, however, is not possible to predict right now.

The bottom line

- Prior to mid-2008, cars and light trucks used more than 90% of transportation energy in the United States, while trains and buses together used only 3%.

- Energy use for transportation is the most easily and quickly changed and therefore is key to rapid improvement in energy conservation.

- U.S. energy use could drop more than 6% if we simply stopped moving coal around to generate electricity. This would offer a double savings, eliminating pollution from the dirtiest fossil fuel and increasing the nation's energy efficiency.

- American society has never decided whether transportation is a public service and thus should be funded by government, or just another commodity that should fend for itself in a free market. America has to make this choice.

- If railroads replace cars and light trucks, energy use for transportation could drop by two-thirds.

- While restoring the infrastructure of the United States will cost $1.6 trillion, new railroads could be built along the major transportation routes in the United States for about $14 billion, less than 1% of the infrastructure-restoration costs.

- Government subsidies for highways cost hundreds of billions of dollars a year, while Amtrak gets less than $1 billion, along with a slap on the wrist.

- Rising gasoline prices have led to a rapid decline in automobile travel, suggesting that in this case the free market worked, but government subsidies have an opposite effect, promoting highway travel.

- The redesign and restoration of major cities can greatly reduce automobile traffic, improve the quality of life in cities, and be an important part of the long-term solution to energy use.

12

Saving energy at home and finding energy at your feet

FIGURE 12.1 Indian dwellings in Mesa Verde, Arizona, showed that the Anasazi understood the benefits of energy conservation, as did all early peoples. They had no choice—they lacked the abundant, cheap energy that we are accustomed to. The Anasazi in Mesa Verde built their houses on the south-facing slopes of cliffs, beneath an arch, so that they were shaded from the intense midday sun but warmed by the early-morning and evening sunlight at lower angles. This was only one of many ways that they lived with only small amounts of fuel. *(Photograph by Daniel B. Botkin)*

Key facts

- In many places in the world before the industrial/scientific revolution, homes and workplaces were designed to conserve energy. That changed in the 20th century. The buildings that we now consider standard and normal are actually novel in human history in terms of energy wastefulness.

- Cave dwellers many thousands of years ago made use of the ability of soil and rock to store heat from the sun, and many of the world's peoples still do so today.

- We are rediscovering geothermal energy and realizing that it could be a major and relatively inexpensive energy source.

- Modern construction materials and careful architectural design can reduce energy use in buildings by 60% or more.

- Costs for these energy-efficient buildings appear only slightly higher than typical 20th-century designs.

Energy-efficient buildings

Long experience throughout human history shows without any doubt that thoughtful design of buildings can save large amounts of energy. The book *A Golden Thread: 2500 Years of Solar Architecture and Technology*, by Ken Butti and John Perlin, tells a beautiful and moving story of how the ancients from many cultures—probably all—designed and situated their homes and public buildings to make them as comfortable as their technology allowed and minimized the need for fuels.

You may be surprised to learn that people are not the only ones who build to make good use of solar energy. Huge termite mounds stand tall among the short grasses in the plains and savannas of East Africa. They not only are warmed by the sun but also have ventilation and even a kind of air-conditioning. Air passages extend from the base of the mound to the top. The sun heats the air at the top, causing it to rise, which in turn draws cooler, oxygen-rich air upward from the bottom. This is "design with nature" designed *by* nature (Figure 12.2).

Recent architectural designs using the best of modern materials have led to new buildings that use much less energy than those characteristic of the Industrial Age. Building for energy conservation is often discussed today as if it is a new idea dreamed up in the most recent decades. But if there has been anything new in modern times—"new" in the sense of novel—it's the way buildings were constructed in the developed nations during the 19th and 20th centuries, when fossil fuels were abundant and cheap. The old lessons were forgotten, and houses and commercial buildings were designed without a thought to energy conservation. With uninsulated or poorly insulated walls, ceilings, and floors, they required large wood-burning fireplaces and woodstoves to keep rooms livable.

FIGURE 12.2 The remnants of a large termite mound in the grasslands of Zimbabwe show that these structures take advantage of passive solar energy but also provide good ventilation and a kind of air conditioning. *(Photograph by Daniel B. Botkin)*

I lived for a year in a classic example of this kind of house, a restored early-19th-century farmhouse in Acworth, New Hampshire. The house, situated on a hillside, had a thick skirting of open boards that separated the downslope side of the basement from the elements. In winter, the winds blew right through these and wafted up into the living room. A furnace had been added in the basement, but did not work, so we heated the house with a shallow fireplace in the living room, another in each of the bedrooms, and a wood-burning cookstove in the kitchen. It was a charming house, and it looked inviting on a summer afternoon, but it was far from cozy on a winter's night.

All this is changing rapidly. Today, friends who live near that house are building houses of amazing new materials invented in recent decades. Imaginative architects and engineers have pursued their use, combining these new materials with the old lessons, lost for several centuries, of designing with nature and with the best of modern technology. Like many preindustrial buildings, these new buildings take into consideration where the sun shines, where the soil and rocks protect, where nature provides easy and convenient energy sources and energy storage, where and how vegetation makes buildings warmer, cooler, more pleasant—and a lot cheaper.

This combination of the new and the old has developed today into a sizable business with many projects, ranging from individual homes to housing developments and commercial buildings. There are many books on energy-conserving building designs; here I can only introduce the topic, focusing on the potential energy savings.

Among many recent examples is a house built in Denver, Colorado, by the National Renewable Energy Laboratory (NREL) in cooperation with Habitat for Humanity. The house makes use of active and passive solar energy and highly insulating modern materials (Figures 12.3 and 12.4). Since its construction in 2002, it has been studied by NREL, which has found that "when the energy efficiency features of the home are combined with solar water heating and solar electricity, the home saves about 60% of the total energy that would be used in an identical home built with standard features."[1] The energy-saving features are grouped into passive and active methods.

FIGURE 12.3 A zero-energy house designed by the National Renewable Energy Laboratory and built in cooperation with Habitat for Humanity. Studies of the house by NREL show that it uses 60% less energy than a house built with standard 20th-century materials and methods. *(Courtesy of DOE/NREL. Photo by Pete Beverly)*[2]

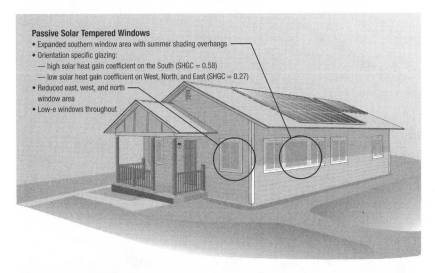

Passive Solar Tempered Windows
• Expanded southern window area with summer shading overhangs
• Orientation specific glazing:
  — high solar heat gain coefficient on the South (SHGC = 0.58)
  — low solar heat gain coefficient on West, North, and East (SHGC = 0.27)
• Reduced east, west, and north window area
• Low-e windows throughout

FIGURE 12.4 Some key features of the zero-energy house designed by the National Renewable Energy Laboratory. *(Reprinted from the National Renewable Energy Laboratory, NREL, 2009. "Zero Energy Homes Research: A Modest Zero Energy Home." http://www.nrel.gov/buildings/zero_energy.html. Accessed December 29, 2009.)*[3]

Passive methods are just that—they arise from the location of a building in relation to the environment, and from the way the building's materials respond to that environment without any additional machinery. Among the passive features of the house are spray foam insulation in the walls, ceilings, and floors; skylights to reduce the need for artificial lighting; less window area on the east and west sides and more window area on the south side; light-colored roof tiles to reduce indoor temperatures in the summer; and adequate attic ventilation. This last feature is crucial; without it, the perfectly insulated building would be a toxic Thermos bottle. The optimum building design allows just enough air flow to provide the oxygen that people need and to prevent the buildup of air pollutants. Some of the cleverest designs have a two-way ventilation system, with air coming in through one set of metal pipes next to another set that allows air to leave. The exiting air can thus pass some of the heat it picked up inside the house to the air coming in, further reducing the need for heating fuels.

Active methods involve machinery to produce electricity and run pumps and other devices to move air and water and to control the use of energy. Active technologies used in the Denver house shown in Figure 12.3 include a 1.8-kilowatt solar-electric system that connects to the electrical grid and feeds energy to the grid when energy production on the rooftop exceeds home use; energy-efficient lighting (in this case compact fluorescent bulbs); and a radiant wall heating system (rather than forced warm air), using heat from a high-efficiency boiler to heat pipes in the walls.

The climate near the ground influences energy use in buildings

In the second half of the 20th century, people began to realize that they could reduce energy use within buildings by taking a new look at some ancient ideas that dominated building design in most civilizations before the availability of cheap fossil fuels. This led to the pursuit of more and better energy-conserving materials. Ecologists studying plants and non-human animals and their relation to their environment began to talk with climatologists and meteorologists about energy exchange, and with architects and landscape planners about human housing.

This has been mutually reinforcing and beneficial. Environmental scientists began to take an interest in local climatic effects and, perhaps most important, in the work of German scientist Rudolf Geiger, who wrote *The Climate Near the Ground,*[4] which showed how very small variations in topography and in the vegetation that covered an area could affect local temperature, humidity, wind speed, and the water content of soil. This work attracted the attention of ecologists, who began in the 1960s to apply its findings to individual organisms, species and their habitats, and ecosystems. Climate was not just "climate, the big picture." There was also *microclimate*—the average weather conditions right around an individual organism that affected the organism and was affected by it. The evolutionary adaptations of animals and plants to the microclimate were astounding and unexpected.

My own first research in ecology was about how trees in a forest adapt to their local climate and how they keep warm enough to survive and grow. One day, trying to place measuring equipment near the top of a small oak tree, I was struck by how different the leaves at the top of this tree were. They were thick, small, and waxy, quite unlike the leaves lower down and those shown in a standard field guide. The treetop leaves were adapted to hot, dry, bright conditions, which afforded all the sunlight they could use but required them to store water. In contrast, the leaves in the shade at the base of the tree were thin, large, porous, and shaped like those pictured in the field guide. (They were easiest to see, draw, and photograph.) These leaves took advantage of the cool, moist surroundings to exchange oxygen and carbon dioxide rapidly and to gather as much light as possible in the heavy shade.

David Gates, a physicist who left that field for ecology, wrote a pathbreaking book, *Energy Exchange in the Biosphere,*[5] that showed how an individual exchanges energy with its local surroundings, including how a person exchanges energy with the house where she lives. He explained that each organism has an energy budget, and he used mathematical equations to predict when a cardinal would have to fly out of the top of a tree because it got too hot and the bird's internal cooling mechanisms could no longer keep its temperature under control.

It wasn't a big step to go from thinking about oaks and cardinals to thinking about one's home. And meanwhile, architects and landscape planners were looking at nature around their buildings and wondering how animals and plants could survive very hot and humid and very cold climates.

We, too, radiate energy and are a source of heat

When we talk about heating a building, we rarely think of our own contribution, but as David Gates explained, we, like all physical objects, exchange energy with our surroundings. We do so in three ways. The first is by giving off heat from our skin directly to the air that touches it (or if the air is warmer than we are, taking up some of that heat). The second is by radiating heat energy—each of us radiates about as much energy as a 100-watt incandescent lightbulb. And the third is by exchanging energy-containing compounds with our environment. The most important of these compounds is water vapor, including what we breathe out and sweat in a warm room.

Whether we feel warm or cool depends on the total energy exchange between ourselves and our local environment—our climate near the ground, so to speak. By the end of the 20th century, standard heating systems blew warm air into a room, air that was not very hot but somewhat warmer than the local environment and warmed a person only if it came in direct contact with the body. But if you stand in a house in the middle of winter in a cold climate, you are also actively exchanging energy with the walls, floors, and ceilings. If they are a lot colder than you are, your body loses out in the energy balance—much energy from your body's surface radiates to the cold walls, floor, and ceiling, and no matter how warm the air from the standard late-20th-century forced-air system, you rarely feel cozy. You can feel hot (and dried out) or chilly (and dried out), but not comfy.

In contrast, in a radiant-heating system, heat is supplied to surfaces, which in turn radiate heat to a room's occupants and also feel warm to the touch. This is more efficient in several ways. I know this from working in the middle of winter in New Hampshire houses that were occupied only in the summer. Somehow nothing felt colder than being in a closed building with very cold walls; it felt warmer outside no matter what the temperature, as far as I was concerned.

I also know this from going into a barn on a dairy farm in the middle of a New Hampshire winter when the cows were inside quietly munching hay or chewing their cud. Because the bacteria digesting woody tissue in the cows' complex stomachs generated a lot of heat energy, the cows were perfect radiant heaters (all the more so if their bodies were black, as a physicist will tell you, because a black surface makes the best radiator). No matter what the weather outside, it was always much cozier in the barn than in the little house I described earlier.

The take-home lesson here is that you feel a lot more comfortable using less energy with a radiant heating system than you do with a forced hot-air system.

But does an energy-smart, comfortable building save money?

In addition to the NREL's formal study of the Habitat for Humanity house, friends and colleagues who have installed the kinds of passive devices just described tell me about amazing savings and increased comfort. One new house in New Hampshire, with modern insulation and high-tech windows, is kept warm, I'm told, with a single small woodstove even though some rooms in the two-story house are quite distant from the stove. In winter, that house uses about one-fifth the wood used by a typical mid-20th-century house of about the same size—two cords of wood compared to ten.

It has already been demonstrated that large energy savings can be achieved through careful design and placement of buildings and the use of the best modern technology. The only question is whether the added costs (if any) of constructing such buildings can be recovered through direct energy savings in a reasonable time, and whether the investment is cost-effective. The answer for retrofitting an existing building will be very different from the answer for a newly constructed building on a lot large enough to allow the best siting in relation to the sun and local topography. Although the general answer may be complicated, NREL states that "energy consumption of new houses can be reduced by as much as 50% with little or no impact on the cost of construction."[6]

The Denver-Boulder, Colorado, area has become one of the major regions for energy-efficient homes, with some companies constructing entire housing developments and others offering to build on an owner's lot. A 2008 article in the *Denver Post* said these houses are being priced 10–15% higher than the equivalent conventional houses.[7] According to the American Institute of Architects (AIA), a comparison of 33 "green" buildings "from across the United States" showed an average cost increase of "less than 2%" versus conventional designs, and savings of $50–65 per square foot, or 20%, arising from energy, emissions, water, operations, and health improvements over a 20-year period."[8, 9] In addition to the features of the NREL-Habitat for Humanity house, some of these new housing developments use geothermal energy for heating and cooling (we discuss this later) and community solar electrical systems to increase energy independence.[10]

Machine Age buildings: the triumph of steel, glass, and cheap energy over human needs

Buildings designed in the industrial/scientific era of the 19th and 20th centuries differed in another important way from most of the shelters people have built during our species' time on the Earth. Buildings came to be viewed primarily as outside of nature, separate from nature, just as people were assumed to be, and just as industry, civilization, towns, and cities were perceived. As such, they were built not just to insulate people from the worst weather and climate, but also to reinforce the triumph of modern technology over nature. Buildings became structures to be admired independent of their surroundings, like the Eiffel Tower in Paris, which demonstrated, with its mathematically determined shape and bare-bones steel structure, that the Machine Age was here, and not just surviving within nature but prevailing. It seemed for a while that the long debate about where people stood in relation to nature was over. With steel and concrete and our fossil-fuel-powered machines, we had prevailed and risen above nature. Nature was no longer our concern—we didn't need it; it didn't matter.

Lewis Mumford, the great 20th-century historian of cities, wrote at length about this aberration in modern designs in such classics as *The City in History* and *The Pentagon of Power*, the latter criticizing the worst of the hubris of the Machine Age. Buildings became objects of art, and by the late 20th century an architect could become famous simply for the beauty of his designs and his drawings, whether or not his buildings worked well for people.

The classic example of this was Yale University's Art and Architecture Building designed by the then head of Yale's School of Architecture, Paul Rudolph. When completed in 1963, the building was praised for being avant-garde, somewhat in the tradition of Frank Lloyd Wright and in the cubist and Bauhaus style. But by the early 1970s it had become infamous. At the time, I was on the faculty of Yale's School of Forestry and Environmental Studies and was invited to participate in a course in the newly fashionable field of ecology at the School of Art and Architecture. The faculty and students told me their woes about the building.

Rudolph had given the architecture students a large work space, with a cathedral ceiling and huge windows. But the drapes were like large fishnets, and their open weave pattern cast constantly moving shadows onto desks and papers, making it difficult for people to focus on their work. He put the sculptors in the basement, where the ceilings were too low for them to do any really large pieces, and where even moderately large stones or statues could be moved in and out of the studio only by an elevator, which was not built for the job and usually ceased to work after transporting a massive stone. This was design-without-nature and design-without-design, and everybody I talked to who worked in the building hated it. The story is told that Rudolph claimed his building was indestructible, built primarily of cement. But much of it was ruined by a fire that became uncontrollable, in part because the stairwells formed open chimneys that allowed the fire to spread.

Green buildings

During the time that steel, glass, concrete, electricity, and modern plumbing, along with cheap and abundant energy, allowed the development of architecture as only art, and of buildings meant to isolate people from their surroundings, another, contrary approach developed. It had its beginnings in the landscape design of Frederick Law Olmsted and in the plans and designs of cities and suburbs by Ebenezer Howard and others. Howard believed that the city and the countryside—that is, the city and its local environment—should be designed together. The idea was to locate garden cities in a set connected by greenbelts, forming a system of countryside and urban landscapes.[11] The names remain, as in Greenbelt, Maryland, and Garden City, New York. Olmsted's use of the natural landscape in designing city parks and Howard's garden city still influence city planning today.[12]

Both Olmsted and Howard saw people as within nature, and saw nature playing an important role even in city life. Olmsted wrote that vegetation in cities played social, psychological, and medical roles; hence, nature within a city was necessary for the best style of life, rather than a city in which the buildings loomed over and dominated people and their surroundings. As Charles Beveridge, the editor of the Frederick Law Olmsted papers and the leading authority today on the history of landscape design, has written:

> The primary purpose of the urban park movement in the 19th century was neither aesthetics nor biological conservation, but was part of a series of sanitary reforms by which those governing cities sought to counteract the threat of ill health produced by industrialization and rapid urbanization. These were pragmatic developments. Among the leaders in the 19th century was Frederick Law Olmsted, who designed New York City's Central Park and had a great deal to say about planning. Olmsted's goal was not aesthetics in itself—he had no interest in beauty for beauty's sake—he was interested in public institutions that met urban psychological and social needs.[13]

The green building idea continues and expands this vision. It takes us beyond a minimalist view of the energy problem—the view that we should all be energy misers, that our only role in nature has been to sin against it, and that we must, like all sinners, pay the price.

Today, we recognize four basic environmental goals for a city: to reduce energy use; to reduce and remove pollutants; to help create a pleasing environment; and to aid in biological conservation. Individual buildings can play a role in each of these—for example, with skyscrapers becoming nesting sites for peregrine falcons in New York City and San Francisco.[14]

Olmsted also thought about individual houses. For example, traveling to Los Angeles, he wrote about how vegetation could benefit those who lived in a city within a semiarid, semidesert environment such as the Los Angeles Basin. "Plantings should be concentrated close to houses, providing an atmosphere of lushness, green and shade, while blocking out the dusty middle distance and setting off distant views so that the dryness and dustiness were not evident, even in a drought season," he wrote. "If such an approach were carried out on hillsides like those at Berkeley, the plantings around the house of one's neighbor below would become a green 'middle distance' in the outlook from one's house, and one's own plantings would do the same for neighbors above."[15]

Research by ecologists helped to stimulate work by the architect Ian McHarg, founder and head of the Department of Landscape Architecture at the University of Pennsylvania, and author of another important book, *Design with Nature*.[16] McHarg became a friend and colleague of some of America's leading ecologists, including Murray Buell, my major professor. In the 1960s, the idea of designing with nature became popular, and among the most famous attempts to accomplish this were Reston, Virginia, and Columbia, Maryland. In the 1970s, with funding by George Mitchell, McHarg designed and built The Woodlands, a suburb of Houston, Texas, where he employed his ideas extensively, focusing on building a connection between people and their environment, so that one's home and town were within a naturalistic setting that minimized negative effects on the environment. The Woodlands has a population of more than 80,000, making it one of the largest towns designed and built in the second half of the 20th century for both people and nature, and for the best interaction between the two.

Today, interest in and actual construction of green buildings has become so much a part of architectural practice that the American Institute of Architects gives annual awards to the ten best green buildings.[17]

In 2008 these ranged from Yale University's Sculpture Building and Gallery to condominiums in various cities[18] and the Cesar Chavez Library in Laveen, Arizona. The Green Building Council, a nonprofit corporation, has established a green building rating system called Leadership in Energy and Environmental Design (LEED) to determine how well a building meets accepted green building conditions. In 2005 the state of Washington passed a green building law requiring all new public buildings greater than 5,000 square feet to meet the LEED standards. The expectation was that energy savings would be 20%.[19]

Energy at your feet: geothermal energy

One of the features advertised for some energy-efficient buildings is the use of geothermal energy—energy from the heat of the Earth. There are two kinds of geothermal energy: the spectacular venting of deep-earth hot gases and liquids, such as the hot springs and geysers at Yellowstone National Park; and low-density, shallow-earth "geoexchange"—the less exciting but for most of us more useful solar energy stored in the earth's soils and rocks, which can be recovered and used.

At present, the first kind of geothermal energy, from deep-earth-heated materials, is used to run steam-electric generators that provide about 7,500 megawatts of generating potential in the United States. This is 7.5% of today's total energy-generating capacity from all the U.S. renewable energy sources, less than half of a percent of the total energy capacity of our nation.[20, 21]

It is the second kind of geothermal energy that is advertised for most energy-efficient buildings.[22] The idea is simple. The Earth's surface—soil, bedrock, and the water stored within these—is warmed by sunlight (and a tiny, tiny bit by deep earth heat generated within the Earth's core). Because these earth materials can store a lot more energy than the atmosphere can near the ground, over time these have become warmer and, just as important, much less variable in their temperature than the air above. If you dig down, you find that as you go deeper, the soil and bedrock very gradually get warmer (in temperate and cold climates), and the temperature varies less from day to night, from season to season, from year to year. That's why sod huts are warmer than cabins built above the prairie (Figure 12.5). It's also one of the reasons why prehistoric people liked to live in caves, such as the famous caves of southern France and Spain, where today we find their 15,000-year-old paintings. Those caves are in limestone hills that are especially good at holding onto the sun's heat.

FIGURE 12.5 Claus Braseth's sod house in North Dakota. A pioneer's sod house on the prairie made good use of passive solar energy and energy conservation. *(Used with permission of the Braseth Family)*[23]

Although the preindustrial way to make use of this geothermal energy was to live in sod houses and caves, the modern way is to put long plastic or metal tubes or hoses down a few feet into the earth and spread them out, sometimes vertically, but more commonly horizontally, sometimes over quite long distances (Figure 12.6). During the winter, the temperature of the soils and rocks is a little warmer than the air; in summer it is slightly cooler. The density of the heat energy stored within the upper surface of the Earth is low, but if you bury long pipes and hoses deeply enough, you can gather a lot of energy.

Some condominium high-rises in Florida make use of this for air-conditioning. One that I know of uses the cool water a few feet below the surface that maintains a pretty steady temperature of about 67°F throughout the year. Instead of having to use a lot of fossil-fuel energy and standard refrigeration equipment to separate cooler air from warmer air, these systems provide air-conditioning simply by passing water or air through pipes cooled by the groundwater, then pumping this air or water through pipes into the apartments. Farther north, doing just the opposite—pumping colder air or water from a wintry surface down into the ground through pipes and hoses—warms the air and water, which then is circulated in a building to provide heat. The only energy we have to expend to get this heating and cooling is for pumping the air or liquid through the circuit of pipes and hoses, which requires much less energy than heating or cooling air or water with a fuel.

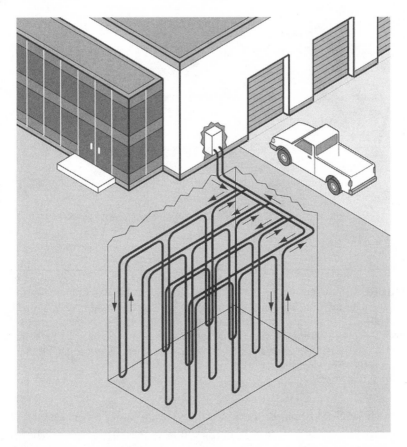

FIGURE 12.6 Heat and cooling right under your feet. How a geoexchange system works.[24] *(Reprinted from National Renewable Energy Laboratory Technical Report, NREL/TP-840-40665)*

Because Earth's soils and rocks are so massive compared with us and our belongings and buildings, vast amounts of geothermal energy exist everywhere on the land (Figure 12.7). This energy is potentially available to us through geoexchange, and indeed an industry is developing to provide geoexchange devices. Proponents say these devices can reduce home heating and air-conditioning bills by as much as 70% and 40%, respectively, although installation costs currently may be about $3,000 more than for a standard air-conditioning and heating system.[25]

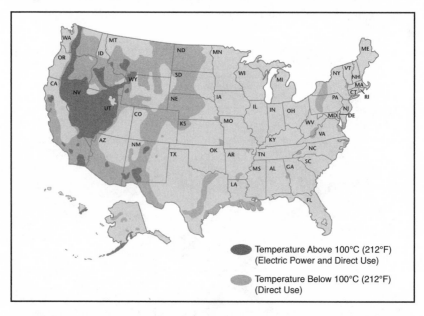

FIGURE 12.7 Geothermal energy in the United States. This map shows the potential for the two kinds of geothermal energy: intense heat from volcanic and other earth activity (dark shaded areas) and geoexchange, which uses the solar energy stored in surface and near-surface rocks and soil (the lightest tone). The darker the color, the higher the temperature. *(Source: National Renewable Energy Laboratory[26])*

The U.S. National Renewable Energy Laboratory estimates that our country could obtain more than 1 million megawatts from geothermal energy, about ten times the amount of energy obtained today from all renewable energy and about 60% of the total energy used today in the

U.S. (Figures 12.7 and 12.8).[27] Remember, this would be a nonpolluting, nongreenhouse-gas-emitting energy source, whose only potential kinds of pollution would come from the manufacturing of the pipes, hoses, and pumps, whatever that might be. However, installing these systems over the entire land area of the United States would pose many problems, including disruption of parks, nature preserves, and cropland, so it is unlikely that their maximum energy potential will ever be realized.

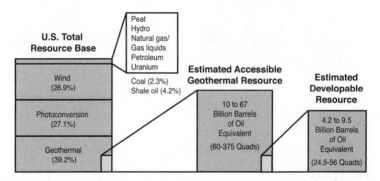

FIGURE 12.8 Wind, solar, and geothermal energy offer vastly greater potential for U.S. energy independence than do fossil fuels, conventional nuclear power, and waterpower. (*Source: National Renewable Energy Laboratory*[28])

According to the NREL, "Today's U.S. geothermal industry is a $2-billion-per-year enterprise involving over 2,800 megawatts of electricity generation capacity, about 620 megawatts of thermal energy capacity in direct-use applications such as indoor heating, greenhouses, food drying, and aquaculture, and over 7,300 megawatts of thermal energy capacity from geothermal heat pumps." The NREL also says that "U.S. geothermal generation annually offsets the emission of 22 million metric tons of carbon dioxide, 200,000 tons of nitrogen oxides, and 110,000 tons of particulate matter from conventional coal-fired plants."

About half of the energy used in a typical American home is for heating and cooling, so a transition to local geothermal could result in a significant reduction in energy use with no change in lifestyle or comfort (Figure 12.9). And of course, this would result in a great decrease in the production of carbon dioxide.

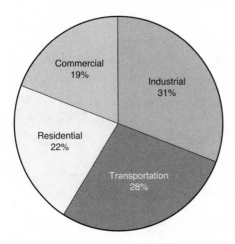

FIGURE 12.9   U.S. energy use by type. *(Source: Energy Information Administration, Annual Energy Review 2008)*[29]

The bottom line

- Energy-efficient buildings and green building designs are major ways that we can reduce per-capita energy consumption and achieve energy independence without sacrificing the quality of our lives. In many cases, the quality of life will likely actually improve.

- Geothermal energy is one of our best bets for energy independence and for inexpensive energy. Geoexchange systems can reduce heating bills as much as 70% and air-conditioning bills as much as 40%.

- Because geothermal energy is locally produced, it also requires fewer expenditures for a national grid, pipelines, or other means of transporting fuels.

- One estimate is that within the lower 48 states this local geothermal energy could provide an energy capacity of a million megawatts.

13

Solutions

All the previous chapters have been leading up to a consideration of what is possible in the future as we transition away from fossil fuels and, in the shorter term, away from petroleum. We've now looked at all the available information about potential energy sources and the advantages and disadvantages of each of them. The tendency has been to champion one source as the complete solution, but the answer is unlikely to be as simple as that. To find our way through the complexities, we look at three scenarios that bracket the possibilities and can help us sort through all of them.

The simple answer to our energy dilemma

The simple answer to our energy problem is for Americans to learn to live happily using just 6% of our current per-capita energy use (the amount Kenyans use). Present installations of nuclear and hydropower would provide all that the U.S. would need in 2050, even with the population increase forecast by the U.S. Census Bureau. We could stop using all fossil fuels, abandon all attempts to develop alternative sources, and not even try to increase the quantity of energy provided by nuclear and hydropower today. But it's unlikely that we could learn to live happily on such a severely restricted energy diet—remember, transportation per capita, and energy use per capita, would have to decline.

That being the case, solving the U.S. energy problem is not going to be simple—it will involve major social, political, and environmental changes.

Is there an answer we can live with—happily?

A solution that would likely be acceptable to most of us would allow us to maintain a high standard of living, perhaps not as high as at present, nor as wasteful of energy, but one that on the whole would seem as good to most of us as it is today. Let's consider several possibilities, focusing on the United States because the data are best; therefore, the necessary points can be made much more clearly and succinctly. It also makes sense to consider possible solutions for the nation that consumes almost one-quarter of the energy used by all the people in the world. For starters, we need to understand the following.

Maintaining both an ample energy supply and energy independence will involve not one energy source but several, and an integrated system that makes the best use of each kind. Not all the energy sources that we use today or are experimenting with today will be major players.

We need a renovated and modernized system to transport energy, through a smart grid and by making liquid fuels from the energy in electricity and transporting it through more and better networks of pipelines.

How to begin

We can't abandon petroleum, natural gas, and coal tomorrow. Alternative energy facilities are presently insufficient, nor will they be up to the task by next year, or the year after. So our first step into our new energy future will be a staged withdrawal from our dependence on petroleum. We can think of 2050 as the deadline for completing our withdrawal from petroleum, because by that year petroleum supplies will be extremely limited and economically impractical if petroleum economists and geologists are correct in their assessments of petroleum reserves. In short, if by 2050 we haven't done something about it, nature will do it for us in its own way.

Let's consider three possible scenarios for the year 2050:

Scenario 1: Business as usual. The U.S. population grows to 420 million by 2050, as currently forecast by the U.S. Census Bureau, while per-capita energy use remains as it is today, as does the percentage of energy supplied by each source.

Scenario 2: Per-capita use as usual. The U.S. population grows to 420 million by 2050, and per-capita use remains the same as today, but the energy comes primarily from solar and wind, largely replacing fossil fuels.

Scenario 3: Alternative energy sources and energy conservation. The U.S. population grows to 420 million by 2050, U.S. per-capita use drops to half the current level (about that of Japan, Great Britain, and Germany), and only a small amount of energy comes from fossil fuel. The question is, which energy source or sources will we have chosen to replace it?

In discussing Scenario 3, we explore largely replacing fossil fuels with solar and wind, as in Scenario 2. Then we consider whether coal, instead of solar and wind, could replace petroleum and natural gas, and explain why nuclear power, ocean power, and natural gas are not viable as the major alternatives.

Of course, mathematically there are an infinite number of scenarios that could be considered. I have selected these three to show the range of costs, as a way to begin to think about the energy future. Economists who read this will quite likely tell you that it is difficult to extrapolate costs into the future, because technological changes affect prices in complex ways. I hope that some economists reading this will be motivated to take what I have written here and improve on the forecasts.

Scenario 1: If America does not change its habits...

Americans need to reduce their per-capita energy use. Table 13.1 and Figure 13.1 show the amount of energy use in the U.S. in 2007.

Table 13.1   U.S. Energy Use in 2007*

Source	Billions of Kwh	Percent
Coal	6,738	23.0%
Oil	11,719	40.0%
Natural Gas	6,738	23.0%
Nuclear	2,344	8.0%
Biofuels	984	3.4%
Hydro	861	2.9%
Geothermal	103	0.4%
Wind	82	0.3%
Solar	21	0.1%
Ocean	—	0.0%
TOTAL	**29,590**	**1.01%**

*Source: Energy Information Agency, U.S. Department of Energy.

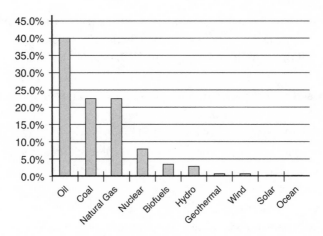

FIGURE 13.1 U.S. energy use in 2007 totaling 29.3 trillion kilowatt-hours.
(Source: Energy Information Agency, U.S. Department of Energy)

According to the U.S. Census Bureau, the population of the United States will reach 420 million by 2050—120 million more people, 40% more than today.[1] If each of the 420 million people, on average, continues to use the same amount of energy that the average American uses today, total energy use would increase from 29 to 40 trillion kilowatt-hours, and the energy supply would have to increase by 40%. Oil would have to provide 16 trillion kilowatt-hours, and coal and gas about 9 trillion kilowatt-hours each[2] (Figure 13.2 and Table 13.2). If the petroleum geologists and economists are correct, that amount of petroleum will not be economically recoverable by 2050. Natural-gas use would have to increase 37%, which is unlikely, period, and especially unlikely without great environmental damage from the methods that will have to be used to mine enough of it. Add to this that hydropower would have to increase 27%, which is highly improbable—as discussed earlier, hydropower is more likely to decline.[3] Conventional nuclear power plants will also not fill the gap in an economically feasible way, if at all, because of the limits of uranium ore.

This is an impossible future. But it is the inevitable future for those who believe we can continue business as usual. Even the near-term future—looking ahead, let's say, only to 2012—is beginning to look grim for maintaining adequate energy supplies if we follow a business-as-usual approach while population and demand grow. According to the nonprofit

North American Electric Reliability Corporation, by 2012 energy demand will exceed supply in most regions of the United States, meaning that the per-capita standard of living is bound to decline without major new investments in energy-generating plants.[4] Major changes in energy supply and construction of major new facilities can no longer be put off in the hope that somehow something will work out "because it always has."

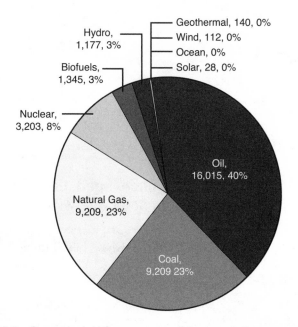

FIGURE 13.2 Scenario 1: U.S. energy use if the population grows by 120 million by 2050 and Americans do not change their habits. Total energy use would increase to 40 trillion kilowatt-hours. According to petroleum geologists and economists, this is an impossible future.

No, our future is not going to be business-as-usual. To maintain a high standard of living, remain a major industrial power, and continue to be a major source of all kinds of creativity—in science, humanities, the arts—we have to move away from petroleum. But, you may say, "we shouldn't have to decrease the amount of energy each of us uses." Okay, then let's suppose nobody has to use less energy, but we do have to move away from fossil fuels.

Table 13.2   Scenario 1: U.S. Energy Use If the Population Grows by 120 Million and Americans by 2050 Do Not Change Their Habits

Energy Source	Amount (Billions of Kilowatt-Hours)	Percent of Total Energy Supplied
Coal	9,209	23.0%
Oil	16,015	40.0%
Natural Gas	9,209	23.0%
Nuclear	3,203	8.0%
Biofuels	1,345	3.4%
Hydro	1,177	2.9%
Geothermal	140	0.4%
Wind	112	0.3%
Solar	28	0.1%
Ocean	—	0.0%
Total	**40,039**	**100.0%**

Scenario 2: Per-capita use unchanged, but reliance changes from fossil fuels to solar and wind

Suppose Americans don't lower their per-capita energy use by 2050 but to a large extent abandon fossil fuels, or what's left of them, and turn to alternative sources of energy. The analysis of all energy sources in the rest of this book can lead you to conclude that the best thing for ourselves and America of 2050 would be a heavy dependence on wind and solar energy. Today those technologies are best prepared to provide abundant energy and the most energy independence, while being the least polluting and best for the environment.

In this scenario of unchanging per-capita energy use, I assume that by 2050 oil, natural gas, and coal will each provide only 1% of the energy in the U.S.—fossil fuels will not be gone and entirely forgotten, but we will have made a planned transition away them, continuing to use them where they are best suited, such as providing energy for peak demand and when wind and solar are putting less into the grid than is required at that time. I also assume that nuclear and freshwater energy will remain at their 2010 amounts, with no net increase or decrease—meaning that the number of hydroelectric power plants remains exactly the same as today, and that the total energy generated by nuclear power plants is the same as today, although some of the plants operating today will have been decommissioned and a few new ones added. As a result, nuclear's contribution drops to 5.9% of the energy supplied, and hydropower drops to 2.9%.

It may be overly optimistic to assume that hydroelectric energy generation will remain at the current level 40 years from now. The U.S. Department of Energy's Energy Information Administration has projected that conventional hydroelectric generation will actually decline 23% by 2020 because of the removal and breaching of more dams for environmental or other reasons.[5] And as I explained in Chapter 4, "Water Power," it is unlikely that in the future any new major hydroelectric power plant will be built in the U.S. (or any technologically developed nation, for that matter).

For this scenario, I also assume that oceans, a yet little-tested energy source, will provide just over 2%, based on the potentials discussed in Chapter 8, "Ocean Power." Ocean energy will begin to play some role, but ocean-energy technology will likely continue to lag solar and wind. A nation intent on expanding ocean-energy use might invest heavily in its research and development, and might boost the contribution above 2% by 2050, but today we have no real basis for planning on that for America.

I also assume that geothermal energy and biofuels will provide 5% each. This is a rather arbitrary amount, based simply on the idea that these two sources will contribute significantly, but not largely, to the total. As explained in Chapter 12, "Saving Energy at Home and Finding Energy at Your Feet," low-intensity geothermal is inexpensive and abundant, and I believe its use is bound to increase.

I have assumed that biofuels will be a minor player for reasons discussed in Chapter 9, "Biofuels." Right now the most promising sources of biofuels are algae and bacteria, but they are still in early development. Crops grown to produce fuels are either net users of energy (energy sinks rather than energy sources), or yield very little more than it took to produce them. As a result, I believe that this technology will be disastrous for our nation, using large amounts of water, straining the supply of phosphate fertilizers, and causing large-scale environmental damage while providing essentially no energy benefit. Despite this, my guess is that lobbyists will not fail completely to get some funding for them. Obtaining energy from waste cooking oils and other wastes is more efficient than not doing so and reduces the amount of new energy required to be generated, but it can never be a major percentage of our nation's energy. We will continue to use firewood, especially where this can be obtained locally.

As a result of these limitations, wind and solar would have to provide 38% each, or more than 15 trillion kilowatt-hours each, to make up the difference (see Table 13.3 and Figure 13.3).

The calculations for Scenario 2 (no change in per-capita use and a heavy reliance on solar and wind energy) are more complicated than for Scenario 1 (business as usual) because of the technological differences between renewable alternative energy sources—wind, solar, and ocean—and energy sources that use a fuel that must be mined to obtain the energy stored within them. With a fossil-fuel-fired electrical power plant, the energy content of a unit of fuel is known, as is the efficiency of energy conversion. In contrast, although each alternative-energy installation has a maximum capacity, the actual yield is a variable depending on environmental conditions.

I have restricted the calculations of solar energy to photovoltaics, not considering solar thermal, for simplicity and because right now it looks like photovoltaics will be the dominant technology. For solar and wind each, the required generation capacity is 15,215 billion kilowatt-hours (Figure 13.3 and Table 13.3).[6] Based on these, in 2050 solar capacity would have to be 12.2 billion kilowatts and wind-energy capacity would have to be 6.48 billion kilowatts.[7] (In the Endnotes, in the section for Chapter 13, you will find a table in note #7 that compares the costs per kilowatt-hour for coal, solar, and wind.)

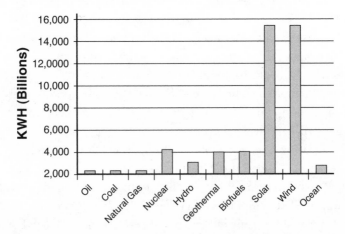

Figure 13.3 Scenario 2: Year 2050, if no change in per-capita use, heavy reliance on wind and solar. The population increases as forecast by the U.S. Census Bureau; coal, oil, and natural gas provide 1% each; nuclear and hydro provide the same quantity as at present; geothermal and biofuels provide 5% each; and the oceans 2%. Solar and wind split the rest and provide 38%.

Table 13.3 Scenario 2: U.S. Energy Use in 2050 if Per-Capita Use Does Not Change but Shifts to Heavy Reliance on Wind and Solar
This table shows the energy production from each energy source if the population increases as forecast by the U.S. Census Bureau, and if coal, oil, and natural gas provide 1% each, nuclear and hydro provide the same quantity as at present, geothermal and biofuels 5% each, and the oceans 2%. Solar and wind split the rest and provide 38% each.

Source	Billions of kWh	Percentage
Coal	400	1.0%
Oil	400	1.0%
Natural Gas	400 -	1.0%
Nuclear	2,344	5.9%
Hydro	1,177	2.9%
Geothermal	2,002	5.0%
Biofuels	2,002--	5.0%
Solar	15,215	38.0%
Wind	15,215	38.0%
Ocean	801	2.0%
Total	**40,039**	**99.8%**

What will this cost?

Estimating the costs of future investments is not simple and it takes us into cost–benefit analysis. Economists agree that in making such estimates, we have to take into account what else we could be doing with the money we are considering investing in something—such as building a new power plant. Economists refer to this as consideration of the "social discount factor." This becomes important in our analysis because economically there are two fundamentally different kinds of energy installations: those that require the continual purchase of fuel (fossil fuels, biofuels, uranium for nuclear power plants); and those that do not (solar, wind, ocean, geothermal). Matthew J. Sobel, William E. Umstattd Professor of Operations Research, Case Western Reserve University, kindly did these calculations that take the social discount factor into account for me, based on the analyses done throughout this book. Here are the results.[8]

Solar energy installations today average $6.81per watt. Wind-energy installations recently have ranged between $1.00 and $3.00 and today average $1.71 per watt. For both, there would be negligible subsequent costs after the initial outlay. (Estimates of maintenance costs are hard to find, but the existing ones suggest they will not be greatly different for coal, solar, and wind.) Using a standard social discount of 5% for Scenario 2, "Per-Capita Use as Usual," solar energy installations would cost $38.6 trillion and wind energy installations would cost $7.6–10 trillion, or a total of $46–49 trillion for the two. If work on these installations began immediately and continued at the same pace until 2050, an investment of about $1.1–1.22 trillion would be required each year in current U.S. dollars.

According to the 2009 federal budget, U.S. federal receipts in 2008 were $2.524 trillion, and 2009 receipts were forecast to be $2,186 trillion.[9] Thus, this business-as-usual approach to per-capita energy use is obviously going to result in an expensive transition from fossil fuels to solar and wind. New technologies and improvements in existing ones will lower the costs, and economists will tell you that the amount of energy people use changes quickly with changes in energy costs. There's also some comfort in knowing that in 2009 the federal budget forecast that 2019 federal tax receipts would rise to $4.446 trillion. Consider also that the 2008 U.S. Department of Defense budget was $0.593 trillion ($593 billion) and the estimated 2009 DOD budget was $728 billion, so the annual transition costs to Scenario 2, with no reduction in per-capita energy use, is 50% higher than the 2009 DOD budget.[10] For those interested in additional comparisons, here are some other funding allocations in the 2009 U.S. federal budget:

- Department of Energy: $26.4 billion
- Transportation: $70.5 billion
- Environmental Protection Agency: $7.8 billion
- Climate Policies (Clean Energy Technologies): $0
- Total for these: $104.7 billion

According to the United Nations Environment Program, there was a total investment of $155 billion worldwide in 2008 in renewable-energy technologies, and "the G-20 group of nations recently announced stimulus packages totaling $3 trillion or 4.5 per cent of their GDP."[11]

These government budget numbers help us to weigh the relative cost of the energy conversion. It's much harder to get an analogous estimate of the total national outlay from private corporations. Clearly, however, there will either have to be massive additional federal expenditures, or the transition will not be funded significantly by the U.S. federal government and will instead have to be done by the private sector. Either way, it will be a major economic and technological commitment.

The bottom line is that Scenario 2, with unchanged per-capita energy use, just doesn't look particularly good.

Scenario 3: Per-capita use drops 50%, solar and wind provide two-thirds

A third and more realistic scenario is of a future in which each person in the United States uses about half as much energy as each American uses today. This means that per-capita use in the United States in 2050 would be about the same as it is today in Great Britain, Germany, and Japan— an energy level at which, on average, these people live well (Table 13.4 and Figure 13.4). The information and analyses in the two preceding chapters indicate that Americans could lower their energy use to this level with little or no loss in the quality of life and, in fact, considering that there would be less pollution and surface mining, probably an improvement.

Table 13.4   Scenario 3: U.S. energy use in 2050 assuming a 50% drop in per-capita use and heavy reliance on solar and wind. Each fossil fuel provides only 1% of the energy; nuclear power provides the same amount it did in 2007 but a greater percentage (11.71%) of the total energy than in 2007. Hydropower also provides the same quantity as in 2007 but a percentage increase from 2.9% in 2007 to 4.3%. Biofuels and ocean energy each provide 5%. This leaves a shortfall, which for the sake of simplicity is accounted for by an increase in low-intensity geothermal (assuming that the costs will be less than that of any fossil fuel energy it replaces).

Source	Billions of KWH	%
Coal	200	1.00%
Oil	200	1.00%
Natural Gas	200	1.00%
Nuclear	2,344	11.71%
Biofuels	1001	5.00%
Hydro	861	4.30%
Geothermal	1,317	6.58%
Wind	6,448	32.21%
Solar	6,448	32.21%
Ocean	1,001	5.00%
Total	**20,020**	**100.00%**

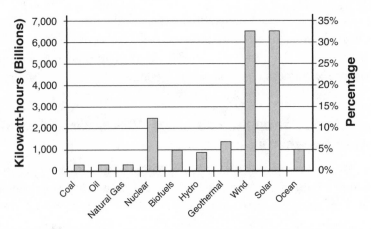

FIGURE 13.4 Scenario 3: U.S. energy use in 2050 assuming a 50% drop in per-capita use and heavy reliance on solar and wind. Each fossil fuel provides only 1% of the energy; nuclear and water power provide the same as in 2007; and biofuels, geothermal, and ocean energy each provide 5%.

Although a 50% reduction seems very large, most of it should be pretty painless for the individual. It will mainly require more-efficient cars, more-efficient cooling and heating, and so forth—technological improvements rather than radical changes in our personal lifestyles. The great advantage is that the demand for solar and wind energy would be only 44% of what it would be if there were no improvement in per-capita energy use. There is one important caveat, however. According to an article in the *New York Times*, "Electricity use from power-hungry gadgets is rising fast all over the world. The fancy new flat-panel televisions everyone has been buying in recent years have turned out to be bigger power hogs than some refrigerators."[12] The International Energy Commission estimates "consumer electronics"—all our computers, cell phones, video games, and so on—use 15% of home energy use, that this is likely to triple by 2030, and, if so, would require "building the equivalent of 560 coal-fired power plants, or 230 nuclear plants."[13] Meanwhile, we lovers of computer gadgets complain about the amount of energy jet airplanes use. Perhaps this is another case of not-in-my-backyard—that we may be looking to solve somebody else's problem far away at 35,000 feet, while ignoring the energy problem occupying our ears and fingertips.

Wind and solar for Scenario 3 are together projected to contribute 6,448 billion kilowatt-hours by 2050 (Table 13.4). Solar energy capacity would have to be 5.22 billion kilowatts, and wind energy capacity would have to be 2.75 billion kilowatts.[14,15] We can ask: Will production capacity for solar and wind meet this challenge? Interestingly, production of photovoltaics has been increasing rapidly, growing by 49% from 2005 to 2006, then another 54% between 2006 and 2007, and increasing a remarkable 91% between 2007 and 2008. Overall, from 1999 to 2008, production increased twelve-fold (1185%), with 987 megawatts produced in 2008.[16] At this growth rate, the goal of 5.22 billion kilowatts required for Scenario 3 would be reached by the year 2037, well short of the deadline year of 2050.

Based on current installation capabilities, the billions of kilowatts of solar capacity required in Scenario 3 would take an area of 5,307 square miles, about 2% of the area of Texas. The 2.88 billion kilowatts of wind turbines would take 1,140 square miles, less than half a percent of the land area of Texas.[17] All the solar and wind energy production for this scenario could be accommodated by about 2.5% of the land area of Texas, or about 0.2% of the land area of the lower 48 states. By comparison, urban area occupies 3% of the lower 48 states, cropland 22%.[18]

Taking into account a 5% annual social-discount factor, here are the results for Scenarios 2 and 3:

Scenario 2 (Per-Capita Use as Usual): Wind and solar replace fossil fuels and together provide 64.4% of the energy required: **Cost**: $91.34 trillion

Scenario 3 (With Energy Conservation): Wind and solar share equally and provide 64.4% of the energy required: **Cost**: $34.77 trillion[19]

This is expensive, but how does it compare to alternatives? Perhaps surprisingly, importing petroleum costs a sizable fraction of the projected installation costs for Scenario 3. For example, in 2008 the United States imported 3.57 billion barrels of oil at an average daily price of $95.62 per barrel, for a total import cost of $341.46 billion. (At this writing in 2010, oil prices exceed $70 per barrel, which would amount to $250 billion annually.) The 2008 total cost of importing oil was about 37% of the annual cost of Option 3's transition to solar and wind by 2050.

Can we plan a reasonable future based on coal?

Coal is the one fossil fuel we're not going to run out of in the next 50 years or so, even using existing technology, and it has been fairly cheap, so it's reasonable to ask: What if Americans opted for a coal-energy future rather than solar and/or wind? But consider: With wind and solar, most costs—close to all from this long-term perspective—are for installation alone, because the energy from then on is free. With coal, after paying to install a coal-fired plant, there are annual costs for the coal itself, all the indirect costs of mining's toxic pollution and destructive effects on the land, and additional costs of "clean-coal" plants to bury carbon dioxide. To make matters even more complex, the National Renewable Energy Laboratory views the total money spent to run a coal-fired power plant (including pollution and land restoration costs) as an economic benefit to the state where the plant is located, while my analysis sees these as expenses.

In June 2008, the U.S, Energy Information Agency recalculated the costs to mine coal and determined that at $10.50/ton, the cost a few years ago, only 6% of the coal in Wyoming, the country's largest reserve, would be economically recoverable.[20] At the time of this writing in 2010, the price of coal delivered to U.S. power plants averages $36.06 per ton. But worldwide, coal is selling at much higher prices, and prices have been rising to as much as $120 a ton (Figure 13.5).

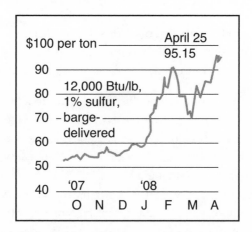

Figure 13.5 The price of coal has been rising rapidly in recent years, for example, doubling between October 2007 and April 2008. *(Source: AP Images/Platts, AP)*[21, 22]

Factoring in a social discount of 5%, the cost of a transition that has coal replacing petroleum and natural gas and providing 64.4% of the energy by 2050 is $31.07 trillion. This is based on the traditional estimates of the costs of building coal-fixed power plants—between $1 and $2 per watt. But as noted in Chapter 3, these costs have been rising rapidly, and one report estimates the cost as high as $3.50 per watt. This would obviously make the total cost much greater.

Could wind do it alone without solar?

We might in theory consider a future in which wind provides 64% of the energy and solar none, in the hope that this would lower the costs even more. Obviously, at present efficiencies and prices, wind is much cheaper than solar, so it would make economic sense to emphasize wind over solar—except for one thing: There is more solar energy available, and it is more consistent.

Interestingly, if wind alone provides 64.4% of our energy needs, the cost falls to $6.44 trillion, which, distributed evenly over 40 years, would be $161 billion a year. (All the cost estimates are summarized in Table 13.5.)

Table 13.5   Total Costs of a Transition to Wind and Solar from 2010 to 2050 (Taking into Account a 5% Annual Social Discount Factor)

For Scenario 2 (Per-Capita Use as Usual) Wind and solar share equally and replace fossil fuels, together providing 64.4% of the energy required:	$91.34 trillion
For Scenario 3 (with Energy Conservation) If wind and solar share equally and replace fossil fuels, together providing 64.4% of the energy required:	$34.77 trillion
If coal replaces petroleum and natural gas and increases to provide 64.4% of the energy (assuming coal-fixed power plants' installation costs remain between $1 and $2 per watt, a low estimate):	$31.07 trillion
If wind alone provides 64.4% of the energy:	$6.44 trillion

The American Wind Energy Association estimates that the windiest 20 states have enough wind-energy potential to provide one-third to one-half of all the energy Americans currently use—and half of the total energy used in Scenario 3. However, with present technologies, depending totally on land-based wind installations may not be feasible in the

United States because of various kinds of opposition to local wind turbine installations, environmental and social. Even allowing for the possibility that additional offshore sites could be found and developed at the
same costs as onshore sites, it's unlikely that a nation would want to go
completely to wind energy, for several reasons, including landscape
beauty, bird mortality, and the risks of relying on just one energy source.
Variety provides redundancy, so if anything goes wrong on a large scale
with one form of energy production, others are available.

In sum, wind is already more cost-effective than coal. The net present value for wind is less than for coal even if coal were free. In fact, coal
would need a \$177-per-ton subsidy to cost as little as wind alone, taking a
5% annual social discount factor into account. Solar, on the other hand, is
unlikely to be cost-effective against coal as both are priced today within
the United States, unless we take into account all the costs associated
with mining and strip mining (costs of erosion, land restoration and
conservation, sedimentation, and health care), which I have not done.
(However, this story could change considerably if installation costs of
coal-fixed power plants triple or quadruple, as some reports indicate.)

As for solar energy, assuming the total costs are in installation, and
there are no maintenance or other costs, solar matches coal when coal
reaches \$433.64 a ton (taking into account net present value).

How about a nuclear future?

I haven't considered nuclear power as the major replacement for fossil
fuel because, as you saw in Chapter 5, "Nuclear Power," with continued
competition for fuel for conventional nuclear reactors, there just won't
be enough uranium. Despite what you may hear from corporations in
France and elsewhere, breeder reactors and nuclear-fuel recycling are
still too experimental, are unlikely to be successful and safe on a large
scale, and are even less likely to provide the large amount of energy
needed in 40 years.

Constructing a conventional nuclear power plant costs \$5–14
billion.[23] For nuclear energy to replace wind and solar in Scenario 3, the
U.S. would need 572 functioning nuclear power plants by 2050—that's
468 in addition to the 104 running now. That would require 12 new
plants a year (the first 12 to be added by the end of 2012). This seems out
of the question, given how much time it takes to determine the plant

design, find a suitable site, get all the approvals, and carry out the construction. It also seems unlikely that sites could be found that would be politically acceptable for 468 new plants. After all, that's an average of more than 10 new plants per state, if the lower 48 states are included. And we could forget about Rhode Island, Delaware, Connecticut, and probably Vermont and New Hampshire, because of their small size, leaving 43 states and an average of more than 11 new plants per state.[24,25]

As discussed in Chapter 5, the lifetime of a nuclear power plant is about 30 to 40 years, and the cost to decommission and dismantle a nuclear power plant is estimated to range from $200 million to more than $600 million.[26] For example, the Maine Yankee nuclear power plant near Portland, ME, one of the first commercial nuclear power plants in the country and one of the first to be decommissioned, cost $231 million to build and is estimated to cost $635 million to dismantle. (And this is an estimate made in 2003.) According to Matthew Wald in a *Scientific American* article, decommissioning this power plant is "an unglamorous task that was not fully thought through during the era when plants were being constructed.[27] Except for one or two experimental power plants, none of the decommissioned nuclear plants in the United States have actually been taken apart and all the radioactive material transported to safe storage.

If we suppose, conservatively, that only the 104 plants that currently exist will have to be decommissioned and dismantled by 2050, that could add as much as a $66 billion to the cost of nuclear power. It is unclear whether these estimated costs are realistic or whether the dismantling cost is going to be in proportion to the construction cost. Because construction costs have gone from several hundred million to more than $10 billion, the dismantling costs could also be orders of magnitude greater than presently estimated. And this assumes no costs associated with accidents or the transportation and storage of radioactive wastes.[28] Nor does it include insurance, because—and this is important—no private insurance company will write a policy for a nuclear power plant, meaning that the federal government is acting as the insurer of last resort, providing another subsidy for nuclear energy.

In sum, no matter what you may have heard, nuclear power would be really, really expensive, in addition to being an environmental and health problem of huge proportions.

However, many proponents of nuclear power suggest building a lot fewer than 468 nuclear power plants, so let's consider what they might contribute. Senator Lamar Alexander of Tennessee has proposed building 100 nuclear power plants in 20 years.[29] If they were built over the next 40 years instead, that would be more than two a year. They would add about the same capacity as all the presently operating plants, providing approximately 2,344 billion kwh a year. This would be about 12% of the energy required in Scenario 3, so the100 new plants plus the existing 104 would provide almost 24% of the nation's energy.

As mentioned, the cost of a nuclear power plant is estimated at $6.8 to $14 billion. The average output is 22.5 billion kWh. Wind turbines to provide the same amount of energy would cost $9.5 billion based on today's average installation costs and energy output. Thus, installation costs of wind turbines appear to be within the price range for nuclear power plants. However, because of a tendency to underestimate their building costs, the nuclear plants can be considerably more expensive to build, and then there's the annual cost of fuel, the decommissioning costs, and any other costs. So it appears unlikely that nuclear power would be cost-competitive with either wind or coal (Figure 13.6).

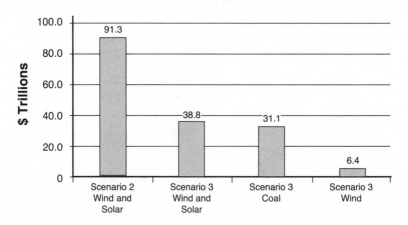

FIGURE 13.6 Comparison of annual costs to convert to non-fossil fuels (using a social discount factor of 5%).

What about natural gas?

I have not considered a natural-gas future because, first of all, as shown in Chapter 2, "Natural Gas," at current rates of use, the world's known reserves of natural gas accessible with known technology will last a short time—about four years according to the latest U.S. Geological Survey information, if America becomes energy-independent; approximately 60 to 65 years, to about 2070, if world resources are purchased—not much longer than petroleum. And the nonconventional sources, right now, are both highly speculative and highly polluting at the source.

So what's the best choice?

In sum, considering only the costs to build power plants, pay for fuel, and deal with coal pollution, wind is the cheapest alternative, coal probably next, and nuclear probably third (assuming there would be sufficient uranium to fuel the power plants). Using solar and wind together to contribute a total of 65% of the energy would be the most expensive. For reasons explained in the chapters on coal and nuclear power, neither coal nor nuclear is a good choice for the environment and for human health.[30]

How to reduce per-capita energy use in the United States

The only workable scenarios for the future—Scenario 3 and perhaps some of its variations—are based on the assumption that we can and will reduce per-capita energy use by about 50%. How can we do this as painlessly as possible? The major ways are mainly technological, not personal: increased energy efficiency in transportation, especially automobiles; increased efficiency of space heating and cooling; and increased efficiency in lighting.

According to the Department of Energy, of the 29,297 billion kilowatt-hours that the United States uses in a year, 28% is used in transportation; 21% is residential (space heating and cooling, lighting, cooking, and running refrigerators and other appliances); 18% is commercial (the same kinds of uses as residential but in businesses of all kinds, as well as all governments and private and public organizations); and 32% is industrial (agriculture and forestry; fishing and hunting; mining, including oil and gas extraction; and construction).[31,32]

Energy used in transportation can be reduced even more than 30% by (1) no longer having to transport as much coal, (2) halving the number of miles driven per person, and (3) increasing the average automobile fuel efficiency to 50 miles per gallon. But it is worth repeating the caveat that the rapid increase in consumer electronics is leading to a large increase in domestic use of electricity, and, if it continues at the present rate, will work against energy conservation.

Storing and transporting solar and wind energy, and using it to transport us

Solar and wind energy are not generated 24 hours a day, day in and day out. Thus, for them to meet a large percentage of our energy needs in the future will require finding ways to store and transport their energy, and ways to convert it into a form that can transport us. Little talked about, but I think necessary to the transition from abundant petroleum and natural gas to alternative energy sources, is the development of large-scale chemical conversions to make liquid fuels from energy generated as electricity.

The basic idea is simple: An electric current passed through water separates the water into its components: hydrogen and oxygen. Hydrogen can be burned as a fuel but is difficult to store and ship. Professor Nathan S. Lewis, of Caltech's Division of Chemistry and Chemical Engineering, has pointed out that chemists can make methane from hydrogen and carbon atoms. When methane is available, the next step is a straightforward chemical process to convert methane to wood alcohol (methanol, whose molecule is just a methane molecule with one oxygen atom added to it) or to ethanol, a slightly more complex molecule.[33,34] Alcohol can be, and is, used to fuel automobiles, trucks, and many other internal combustion engines.

Some more imaginative chemistry can then take hydrogen, methane, ethanol, and methanol and make gasoline and jet fuel in a kind of reverse refinery process. Each chemical step uses some energy, so the overall energy efficiency goes down, but there is enough alternative energy for this not to be a serious problem. Now is a good time for the big chemical, petrochemical, and power corporations to get together and start developing the chemical technology and building the reverse refineries that will be required.

Improving transportation energy efficiency

If Americans do not drive less, and automakers do not increase the average miles per gallon of fuel, then in 2050 the United States will need

173 billion gallons of gasoline. But we've already proved that we can manage very well with less. In response to the soaring prices of gasoline in 2008, people stopped buying gas-guzzlers and pickup trucks, used public transportation more, and took shorter vacation trips or none at all—taking "staycations" at home and pretending to be on a trip. The Energy Independence and Security Act of 2007 requires that passenger cars average 35 mpg by 2020, a 44% improvement overall but still asking for a less than 1% improvement in gas mileage per year.[35] If the miles driven per person dropped 50%, the total fuel used by motor vehicles in the U.S. would drop to 42 billion gallons, 10% less than we used in 2007.

More than 6% of all energy used in the United States is for transporting coal, so if the nation stopped burning coal, U.S. energy use for transportation would decline by that amount. In Scenario 3, energy from coal declines from 6.7 trillion kilowatt-hours to 205 billion kilowatt-hours (just 3% of what it was before).

The costs to build railroad lines are said to be no higher than $2.5 million a mile for construction and equipment. Taking into account the additional costs to purchase land and rights-of-way, a reasonable average cost is between $20 million and $40 million per mile. On this basis, the price to build from scratch a new high-speed railway between Los Angeles and San Francisco or Sacramento would be as low as $700 million. Adding in the cost to purchase land would bring the price to between $7 billion and $14 billion.

This is comparatively inexpensive in comparison to the installation costs of new power plants—about the same as one nuclear power plant or a wind turbine installation that provides the same energy output as the nuclear power plant, with a big payoff in energy conserved. It is also small compared with the estimated $1.6 trillion needed to restore America's infrastructure—including bridges, tunnels, highways, airports, sewage lines, dams, hazardous-waste disposal, schools, and navigable waterways.[36]

So, why all the opposition to high-speed rail in the United States? The claim is that it is too costly without adequate payback, but if in restoring America's infrastructure we shifted our emphasis from highways to railroads, the costs would be small by comparison to everything else we have discussed. This leaves us with the sinking feeling that the opposition to railroads cannot be based simply on overall economics but instead is influenced by special interests in automobile and truck transportation, the building and repair of those vehicles and their highways, and the profits made from selling fossil fuels to run them. And without question it is also influenced by a cultural attitude about railroads, that they are just plain old-fashioned and for that reason alone not worth bothering with. Add to this the wonderful convenience of the personal automobile and you have an arsenal with which to argue against railroads, as if it were an economic argument, when it isn't.

Savings in residential, industrial, and commercial use of energy

Savings in these areas could be accomplished by using low-density geothermal energy and passive solar energy—that is, the natural flow of energy without mechanical pumps—to move air or water and by using modern insulation and insulated window glass in heating and cooling, as discussed in Chapter 12, "Saving Energy at Home and Finding Energy at Your Feet."

Solving the energy problem the American way, which is ...?

If the third scenario—reducing our per-capita energy use by 50% and replacing fossil fuels with wind and solar—is the future, how might it be achieved in the United States? No doubt there will be heated debate as to whether the changes in energy sources and energy use would best be achieved by the free market or by the government. But the answer seems clear from the history of photovoltaic manufacturing in the United States. At the current rate of increase in photovoltaic production, the amount needed for Scenario 3 would be reached by 2037 if all photovoltaics produced in the U.S. were installed in the U.S. Whether the future manufacture and installation of solar facilities will be determined by the free market or by government depends largely on whether our society views the supply of energy as a social service, therefore to be funded by government, or as just another commodity to be traded in a free market.

The modern world has not settled this question, one that is especially troubling and is at the heart of the energy debate and energy dilemma of the United States. If energy is just another commodity, then an argument can be made that it is not the business of government, but only the business of business.

Mulling this over, I called Tom Veblen, a retired corporate executive whose career included major positions with Cargill, a corporation that describes itself as "an international provider of food, agricultural and risk-management products and services." Tom organized and for many years has led "The Superior Business Firm Roundtable," an informal gathering of retired CEOs and other interested people who discuss what makes for a superior business, what are the roles of business in a democratic society, and, by extension, what makes for a better society. I told Tom about the conclusions I had reached and asked how he thought America should approach solving its energy-supply problem.

As a believer in free-market capitalism, Tom replied that we should do nothing, that the market would take care of it, that the rising price of energy would drive activities to conserve energy. But, he said, this will never happen, because elected officials will be asked to do something and will feel that they have to do something. In that situation, he suggested, the government should promote mass transportation and the rebuilding of inner cities to reduce overall energy consumption.

Those with different economic/philosophical perspectives will point to Nick Taylor's 2008 book *American-Made: The Enduring Legacy of the WPA*.[37] The author reviews the events of the Great Depression as a time when the free-market philosophy per se failed. Before becoming president, Herbert Hoover, an accomplished and smart man, had risen to fame when he was secretary of commerce and led a response to the 1927 Mississippi River flood. He had gone on the radio and helped the Red Cross raise $15 million for flood victims. He had visited 91 communities suffering from the flood, and coordinated eight government agencies, arranged for 600 ships with supplies and a trainload of feed for cattle. Altogether, he appeared to be a man who cared for the welfare of the ordinary person.[38]

After becoming president, however, he said no to a proposed federal response to the economic crisis, believing that business is not the role of government, and that helping the needy was the province of private charities. In 1932 he vetoed a $2 billion public-works jobs plan, which he called "a squandering of public money," while thousands of out-of-work veterans of World War I marched on Washington seeking jobs. When the unemployed veterans camped in Washington, Hoover sent in troops led by Douglas MacArthur and George S. Patton. The campsite was destroyed, and several small children were killed.[39]

When Franklin D. Roosevelt became president, he appointed Harry Hopkins to head a program under the newly passed Federal Emergency

Relief Act. In some states 40% of the people were out of work, and in some counties 90% of the people were on relief, and people began to agitate for improvements. Roosevelt pushed legislation establishing the Tennessee Valley Authority that built major hydroelectric dams in the Southeast, and later the Bonneville Power Administration to do the same in the Pacific Northwest, the point being not just to generate electricity but to provide jobs. Hopkins meanwhile helped the Civil Conservation Corps get started on employing out-of-work people in doing good works: building fire towers, clearing firebreaks in forests, stocking fish in rivers, planting trees to retard erosion, and so on.[40] Roosevelt got the National Industrial Recovery Act passed, which allocated $3.3 billion for dams, bridges, and other large projects.

Proponents of this kind of government action say there are times when the free-market approach just doesn't work, at least not fast enough to avert a crisis and ease the suffering of millions of Americans. They firmly believe that today's energy crisis is one of those times. They fear that the rapid increase in the cost of energy and decline in energy availability will have disastrous effects on the American economy and possibly help to cause a major depression. In sum, they argue that at least temporarily federal programs will be necessary to promote the transition to alternative energy. A Civil Energy Corps, for example, could employ out-of work-people to build modern intercity and intracity railroads, retrofit government buildings for more efficient heating and cooling, and so forth.

My own experience during the last ten years has brought me face-to-face with a curious contradiction: Although the U.S. government provides large subsidies for fossil fuels and nuclear energy, even my environmentalist and environmental-economist friends and colleagues almost always raise the question of why solar and wind can't pay for themselves. It's interesting, and disturbing, that the old energy sources that pollute the environment and are going to run out soon will continue to benefit from special-interest lobbying and heavy government subsidies, whereas developing new forms of energy to replace them must pay for itself. In my view, it's not really a "free market" unless either all energy sources are equally subsidized or none are subsidized. But this is unlikely to happen at present because of the heavy influence of special interests.

My guess is that most Americans will see energy not as a completely free-market commodity but as a combination of social service and commodity. For example, we are currently seeing what happens when a major form of transportation, air travel, is treated not as a national and international necessity but as an ordinary commodity whose problems are left to be solved by the marketplace. Regulated only minimally now, primarily for safety and traffic flow, airlines are having difficulty making a profit and are cutting flights to all but the most profitable destinations. Some existing airlines may not survive. Because air (and rail) transportation is essential to our market economy and fundamental to our way of life, most Americans will almost certainly want some government assurance that adequate transportation to and from major cities will be restored and maintained.

Many years of interactions with government agencies that deal with natural resources, planning, scientific research, and data gathering have left me with a mixed picture and mixed feelings. Although it has become clear to me over the years that huge government bureaucracies are inefficient and often unproductive, these agencies have nevertheless done some necessary things that would not have happened otherwise. Our highways, railroads, and hydroelectric dams are just a few examples.

All in all, I'm convinced that without some social/economic incentives, the transition from fossil fuels to alternative fuels is unlikely to happen in time to avoid a full-blown crisis. As I read the newspaper every day, I see things that government and private enterprise need to do now, things they need to do soon, and things that need to be done in the future to help us accomplish the transition to new energy systems. Here is a list of some of those things.

Proposed energy program for federal and local governments

Build federally funded solar and wind plants to produce gas and liquid fuels. Within the next year, the Department of Energy should build a 10-MW solar energy plant and a 10-MW wind energy plant whose output is used only to produce gaseous and liquid fuels, as a start on the technological development that will be needed in the future.

Put a similar installation on a military base. In addition to the 2009 Department of Defense decision to fund research for algae-produced renewable F-76 Naval Distillate fuel (which is a good development and probably should expand), the Department of Defense should build a solar/wind facility of similar size on a military base to produce liquid fuels for military vehicles. Both Department of Energy and DOD facilities would have a number of small gas turbines—aircraft jet engines slightly modified to generate electricity from liquid fuels to test the new fuels. (Note: These are already in use to meet peak power demand at many locations.)

Use land on military bases for wind and solar installations to supply energy beyond the bases. Many military bases have large land areas that serve as buffers from the rest of the country for the safety of citizens. Some of these large tracts are available for biological conservation and could be used for wind and solar energy generation. This has been discussed for several decades for Vandenberg Air Force Base in Southern California.

Increase federal funding for alternative-energy research and development. U.S. political leaders currently propose spending from zero to $10 billion a year for alternative-energy research and development. The president and Congress should greatly increase this amount.

Level the subsidy playing field, by either eliminating all subsidies (probably a political impossibility) or equalizing subsidies (also a politically difficult feat). Ideally, eliminate all subsidies for oil and gas, and use that money instead to fund alternative energy. Or revise the conditions of these subsidies so that the oil and gas companies use them to develop liquid fuels from alternative-fuel electricity.

Encourage development of ocean energy. Private funds, from foundations and individuals, should be used to establish a series of prizes for technological development of ocean energy and for vehicles that use forms of alternative energy.

Popularize vehicles that use alternative energy. Establish an automobile race like Australia's solar-energy-car race but in this case including vehicles powered by wind, solar, or ocean energy that has been converted to electricity or a liquid fuel. This will attract attention to research and development of such vehicles and boost their popularity.

Offer a prize for the first rocket to launch a satellite into Earth's orbit using liquid and gas fuel produced from electricity generated by wind and solar energy.

Enhance building codes in cities and towns to require improved efficiency of space heating and cooling and require the new highly insulating windows.

Refit government buildings to save energy, using the best insulation and windows to improve efficiency of space heating and cooling.

Reduce the need for private cars in cities, not by punitively taxing the use of cars but by improving public transportation, making walking and bicycling safer and more attractive, and promoting such things as ZIP cars, and concentrating development around "intermodal transportation" hubs—locations where several kinds of transportation come together. Much of the technology to do this exists now.

Embark on a major program to improve rail travel.
For example:

- Provide railroad subsidies equal to highway subsidies.
- Provide high-speed rail from Boston to Washington; Washington to Atlanta; Los Angeles to San Francisco; San Francisco to Seattle. To do this, revise the EPA and OSHA rules and other rules so that French, German, or Japanese train technology can be used directly in the United States, rather than the unnecessarily heavy and perhaps unworkable current Acela designs.
- Revise railroad rules to give passenger trains the right-of-way when both they and freight trains use the same tracks.
- Improve the nation's train tracks and roadbeds.
- Link major airports of major cities to the central city by high-speed rail.

Boost employment and help conserve energy during economic downturns. When many people are out of work, the federal government should establish the Civil Energy Corps, which would

- Employ people to retrofit government buildings and private residences with energy-saving materials such as wall insulation, weather stripping, and insulated windows;
- Employ landscape architects to plan, design, and develop wind and solar facilities that not only generate energy but also improve landscape aesthetics.

Improve public access to energy information. For example, the Department of Energy says that it is no longer reporting the costs of dealing with nuclear waste. Reinstate public access to this vital information.

Major conclusions

What won't work

Natural gas cannot make America energy-independent without major environmental damage—and even with it, available reserves may be insufficient.

- Within U.S. lands and waters, natural gas accessible by known technology will last only a year or a few years; the rest is in methyl hydrates (which lie in the deep ocean or in permafrost), coal gas, and shale. The technology to mine these is in limited development, and its success remains unknown.

Conventional nuclear power plants cannot make America energy-independent at a reasonable cost. The supply of uranium ore is too limited, and as world demand for it increases, the price will rise rapidly. At present, nuclear power plants are also among the more expensive to build.

- To replace the use of fossil fuels in the U.S., 468 new nuclear power plants will be needed by 2050, almost ten per state. This is just not likely to happen.

- Nuclear power plants have a fixed lifetime, after which they have to be dismantled and their radioactive wastes stored and protected for a long time.

- Contrary to assertions by nuclear-power enthusiasts, dealing with radioactive wastes from nuclear power plants remains an unsolved problem in the United States.

- Except for one or two experimental reactors, no commercial nuclear power plants have yet been completely dismantled.

Nonconventional nuclear power plants are not yet ready for prime time—breeder reactors, large-scale recycling of fuel for conventional nuclear reactors, and fusion reactors are still in the research-and-development stage for large-scale and widespread installations.

- Fusion has yet to prove technically possible after a half-century of research and attempted development

- There are few operating breeder reactors in the world, and it is unclear which and how many are actually "breeding" in an environmentally benign and cost-effective way.

- Little recycling of spent conventional nuclear fuel is being done, and it is unclear whether this will be technologically possible in ways that are environmentally benign and cost-effective.

Conventional water power—large dams and reservoirs on major rivers—cannot increase in any significant way and is likely to decrease in the United States in the next 40 years.

- All the good U.S. sites are already in use.
- Many dams are likely to be breached or removed because of environmental concerns.

Deep-earth geothermal energy—from volcanically active areas, such as Hawaii and Yellowstone National Park—can be only a minor contributor to America's energy. Not enough can be made available, and there are major environmental and cultural reasons not to go this route.

Agrifuels—land crops grown to produce fuels, not food—are among the worst sources of energy. In most cases, they take more energy to produce than they yield, and even if they are slightly energy-positive, they cause great environmental damage and take land, water, and fertilizers away from food production. This is especially an economic problem for phosphorus fertilizers. In some places, land that provides habitat for threatened and endangered species or is useful in many other ways is being converted to growing fuel crops. *All in all, agrifuels should be avoided.*

What's questionable

Ocean power offers some potential. Even though the most optimistic estimates suggest that this cannot be the key to America's energy independence, still, research and development should be supported and expanded.

Coal is sufficiently abundant to make America energy-independent, but, contrary to conventional wisdom, will not be a cheap alternative. Taking into account the 5% annual social discount factor, plus purchases of coal for fuel, and the costs of pollution impacts, coal is much more expensive than wind. In fact, for coal to be cost-competitive with wind would require a government subsidy exceeding $100 a ton—the cost of building coal-fired power plants, which appears to be rising rapidly.

What will work

Nonconventional water power—such as submerged turbines in free-flowing rivers—could make some small contribution, insignificant nationally but perhaps useful locally in certain areas.

Low-level geothermal energy is one of the cheapest and best ways to heat and cool buildings and can be an important contributor to America's energy independence. This is energy that originates from the sun or in some cases from the Earth itself and is stored in soils, rocks, and underground water. For political and cultural, and technological reasons, it's hard to estimate how much energy this can provide in the next 40 years. A major increase in research, development, and installation of existing technology is needed.

Wind today is as cheap as any energy source, and in the future, taking into account net present value, cheaper than any alternative. Wind will be one of the major sources of energy in the future, and its use is increasing rapidly. It can provide a large fraction, but not all, of our energy needs.

Solar energy is the largest and most reliable source, with a potential greatly exceeding what people could ever use in the next centuries, but at present it is expensive. Today's off-the-shelf devices convert 20% of solar energy into electricity. Great scaling up is needed, and the technology is there. However, right now solar energy is much more expensive than wind, coal, or low-level geothermal.

- Like it or not, solar is going to be a major contributor to America's future energy supply. How costly it will be depends on how much research and development goes into it.

- The 2008 total cost of importing oil equals about 37% of the amount required to transition to solar and wind by 2050.

The one great hope for biofuels is microorganisms—algae and bacteria. Some of these can produce methane (natural gas), biodiesel, or ethanol directly, without distilleries and expensive transportation adding to the production cost. But this energy source is only in the exploratory stage. The potential appears great, but as yet, it's impossible to estimate how much these could contribute over the next 40 years.

Using organic wastes as fuels—waste cooking oil, waste-paper—can contribute to our energy efficiency, and it certainly is a more efficient use of wastes than dumping them in the ground.

However, these are not net energy sources; using them recovers some of the energy stored in them.

What else is needed?

Improved energy transport. None of the energy sources—neither fossil fuels nor conventional or unconventional alternatives—can satisfy America's energy needs without major improvements in the way we transport energy. This requires extending the electrical grid and developing a smart grid. It also requires a kind of reverse refinery, where liquid fuels can be produced economically from electricity and then transported by pipelines, railroads, and trucks.

- Further improvements will involve microgrids and land-use plans that bring heavy users of energy closer to energy sources so that energy will have to be transmitted only short distances without using a national grid or pipelines.
- Off-the-grid local installations of wind and solar can contribute to America's energy independence in ways not possible before.

America's energy independence also requires improved and extended energy storage, both for specific devices (better batteries) and for large-scale installations.

Right now, it is unclear whether batteries or liquid fuels will be the best energy source for autos, trucks, and buses in the future. We have become so used to gasoline, kerosene (jet fuel), and diesel that we forget what marvelous energy-storers they are. It's going to be hard to replace them but perhaps not necessary if we learn to make them from algae, bacteria, and electricity-operated reverse refineries.

A final word

In sum, I believe that America needs abundant energy. It is necessary for our leadership in science and technological development, for our culture, arts, humanities, health and welfare, and also appears important to democracy, as energy-poor people need to focus more on survival than on political science. I also believe that Americans can continue to have abundant energy, not by becoming energy misers but with improved energy conservation and the use of energy sources that are environmentally benign and cost-effective. We can achieve this with little or no change in our standard of living or in the quality of our lives. In fact, it could lead to an improvement in both.

I have tried to provide objective information and analysis that citizens of a democracy can use to arrive at their own conclusions about what needs to be done to ensure an ample energy supply for the future. But stepping out of my role as a scientist, I have to say that the implications of the information seem clear, and it would be incorrect to leave the matter without telling you what it means to me and what I believe needs to be done and should be done. For what it's worth, here is my judgment.

Today the energy debate is sometimes expressed as only a subset of the debate about global warming. But we need to move away from fossil fuels for a number of reasons:

1. They are going to run out. Oil will run out soonest, and as we are already seeing, its diminishing availability is causing its price to rise rapidly (which petroleum experts have for years been warning would happen).

2. Moving away from oil and natural gas is in the best interest of America's foreign policy and military security and the safety of its citizens.

3. Petroleum is made up of many compounds that are used in manufacturing all the plastic articles, too numerous to count, that we use every day. Rather than burn it, we should save petroleum for those uses. They require considerably less than what we burn, so the petroleum we have left will last longer.

4. The trade-off for petroleum's wonderful fuel derivatives—gasoline, diesel, and kerosene—is their toxicity and their pollution. We have learned to live with these to some extent, but they are not the best choice for our health and for other living things.

If the diminishing supply of oil and gas weren't forcing us to seek other energy sources, these considerations, and current concerns about global warming, would make a strong case for moving away from petroleum and natural gas as soon as possible.

As for nuclear power, the abundant writings about it, while claiming that it is safe, rarely discuss realistically the potential costs of dealing with radioactive waste and nuclear accidents. Having worked with radioactive materials and in an experimentally radioactive ecosystem, I disagree. I am convinced that it is in the best interests of humanity and civilization that we choose an energy path that minimizes pollution of all kinds and promotes, to the extent that it can, the beauty of landscapes, the diversity of life on Earth, and the livability of our villages, towns, and cities.

Even if you don't agree, let me reiterate that, like fossil fuels, nuclear fuel is in short supply and will likely run out in 40 or 50 years or less, which in itself makes it pointless to invest in infrastructure based on it.

The analyses I put together in writing this book suggest that moving away from fossil fuels and nuclear energy is possible but expensive. I concluded that we should pursue the two most viable alternative sources of energy: solar and wind. We should also pursue the development of energy from the ocean, and seek ways to use all of our energy sources as efficiently as possible. We should do these things because they will be best for our descendants, their civilizations, their creativity, health, welfare, and happiness, and best for all of life on Earth.

Endnotes

Preface

1 Dr. John H. DeYoung, Jr., Chief Scientist, Minerals Information Team, U.S. Geological Survey.

Introduction

1 Unless otherwise noted, the material in quotation marks throughout this section is taken directly from Behr, Peter, and Steven Gray, "Grid Operators Spotted Overloads but Ind. Controllers Couldn't Force Power Companies to Cut Output," *Washington Post*, 5 September 2003, E01.

2 Accuweather records for August 14: high 91, low 75, average 83; normal temperatures for those dates: high 83, low 68, average 76; rain on August 14: zero.

3 U.S. Census Bureau, *2000 Census of Population and Housing, Population and Housing Unit Counts PHC-3-1* (Washington, D.C.: 2004).

4 1 horsepower (HP) = 745 watts, or 0.75KW. A 100HP car engine, therefore, is a 75KW engine. An Olds Cutlass 88 with a 365HP motor had the equivalent of 365 x 0.75 KW = 274KW engine.

5 Behr, Peter, and Steven Gray. "Grid Operators Spotted Overloads but Ind. Controllers Couldn't Force Power Companies to Cut Output," *Washington Post*, 5 September 2003, E01.

6 Federal Energy Regulatory Commission, Open Hearings about the August 14, 2003 Blackout, 2004; and Behr, Peter, and Steven Gray, "Grid Operators Spotted Overloads but Ind. Controllers Couldn't Force Power Companies to Cut Output," *Washington Post*, 5 September 2003, E01.

7 U.S.–Canada Power System Outage Task Force 2004 Final Report on the August 14th Blackout in the United States and Canada, April 2004. https://reports. energy.gov/.

8 Kilpatrick, Kwame M., Mayor, City of Detroit, Federal Energy Regulatory Commission Open Hearings about the August 14, 2003 Blackout, 2004, p. 94.

9 Hanson, Holly, et al., "Everyday Chores Are Test of Ingenuity," *Detroit Free Press*, 16 August 2003.

10 *Ibid.*

11 Renewable Energy Industry website, www.renewable-energy-industry.com/news/newstickerdetail.php?changeLang=en _GB&newsid=1097.

12 Kilpatrick, Kwame M., Mayor, City of Detroit, Open Hearings about the August 14, 2003 Blackout, 2004, p.94.

13 www.seco.cpa.state.tx.us/re_wind.htm, accessed 30 April 2008. Consider some other ways to think about energy. One megawatt (MW) is enough electricity to serve 250–300 homes on average each day. That works out to 3KW–4KW a house, or 30–40 100-watt bulbs burning continually.

14 This section is based on Botkin, D. B., and E. A. Keller, *Environmental Science: Earth as a Living Planet* (New York: John Wiley & Sons, 2009).

15 Butti, K., and J. Perlin, *A Golden Thread: 2,500 Years of Solar Architecture and Technology* (Palo Alto: Cheshire Books, 1980).

Section 1

1 From DOE EIA Table ES1. "Summary Statistics for the United States, 1994 through 2005."

2 Energy is the work done by moving an object of known weight a unit distance. For example, when you work out at the gym and lift 25 pounds in a curl, moving the weight, say, 3 feet, you have done 75 pound-feet of work. Power is energy used (or generated) per unit of time. So if you do one curl in 3 seconds, your power output is 25 pound-feet per second. Power plants are rated in terms of power. The energy output depends on how long they run. Typically, the energy output is written per hour, day, or year.

3 To be precise, annual U.S. energy use is equal to 9,267 100-watt light bulbs burning for each person all the time. Another common unit in which energy is expressed is the British thermal unit (BTU). This is an old-fashioned term: the amount of heat to raise 1 pound of water from 60°F to 61°F (at one standard atmosphere of air pressure). Often discussions of global or national energy use are expressed in BTUs. But since a single BTU is so small, the energy used by a nation or the world is written down as *quads*. A quad is a quadrillion BTUs (a million billion BTUs) or 293 billion kilowatt-hours. In these terms, the United States' energy use is about 100 quads; worldwide, people use 462 quads a year.

4 Consider two more useful numbers: The U.S. has the capacity to produce 978,020MW of electrical energy. Of this, 754,989MW are from fossil fuel power plants. Remember, this is capacity, not actual output. That is, if all these generators ran at full capacity for one hour, they would produce 978,020 kilowatt-hours of electricity. If they did this for a day, they would produce 978,020 × 24 kilowatt-hours. DOE EIA Table ES1. "Summary Statistics for the United States, 1994 through 2005."

Chapter 1

1 Heywood, John B., "Fueling Our Transportation Future," *Scientific American* special issue, "Energy's Future: Beyond Carbon" 295 (2006): 60–63.

2 Bockstoce, J., "On the Development of Whaling in the Western Thule Culture," *Folk* 18 (1976): 41–46; and Bockstoce, personal communication with author, September, 2008.

3 UCSB Geography Department slide presentation, "Introduction to Air Photo Interpretation Slides," no. 7. www.geog.ucsb.edu/~jeff/115a/jack_slides/page7.html.

4 U.S. Energy Information Administration, *U.S. Crude Oil Supply & Disposition* (2007). http://tonto.eia.doe.gov/dnav/pet/pet_sum_crdsnd_adc_mbbl_a.htm. This reports states that the U.S. used 20 million barrels of oil a day, importing 55% of it.

5 Different sources give slightly different values for the amount of total energy and electrical energy provided by petroleum and natural gas.

6 www.energy.gov/energysources/fossilfuels.htm.

7 Heywood, John B., "Fueling Our Transportation Future," *Scientific American* special issue "Energy's Future: Beyond Carbon" 295 (2006): 60–63.

8 EIA, http://tonto.eia.doe.gov/energyexplained/index.cfm?page=electricity_in_the_United_States.

9 Botkin, D. B., and E. A. Keller, *Environmental Science: Earth as a Living Planet* (New York: John Wiley & Sons, 2009).

10 EIA slide presentation, "Long-Term World Oil Supply, 2000." www.eia.doe.gov/pub/oil_gas/petroleum/presentations/2000/long_term_supply/sld 009.htm. Accessed 7 May 2008.

11 Some argue against the idea of peak oil production, among them Michael Lynch, former director for Asian energy and security at the Center for International Studies at the Massachusetts Institute of Technology: (See Lynch, Michael, "'Peak Oil' Is a Waste of Energy," *New York Times*, August 25, 2009.) His argument is simple, and one comes across it often via advocates of petroleum: Experts disagree on how much oil exists and new finds change the estimate, so peak oil is not known with any certainty. Therefore, we should "be happy, don't worry." This is an argument without substance. Any business planner—or careful planner of any kind—would use the mean and variance of estimates of total peak to create a statistical useful estimate of the time to peak oil, which has to happen.

12 Saleri, N. G., "The World Has Plenty of Oil," *Wall Street Journal,* 4 March 2008, A17.

13 *Ibid.*

14 Botkin and Keller, 2009; and British Petroleum Company, *BP Statistical Review of World Energy,* (London: British Petroleum Company, June 2007).

15 British Petroleum Company, 2007.

16 *Ibid.*

17 Gibbon, G. A., U. S. Energy Sources and Consumption PowerPoint presentation, sent to the author as a personal communication, June 2007.

18 Botkin and Keller, *Environmental Science*, 2009.

19 Baskin, Brian, "Northern Exposure: As the Arctic gets warmer, oil and gas producers see the chance for a big expansion. But plenty of technological hurdles remain." *Wall Street Journal*, 11 February 2008, R12.

20 *Ibid.*

21 The forecast Arctic addition to petroleum is an amount that would increase the present known reserves by 40% and, at the rate of use of 50 billion barrels a year, would provide eight years of oil use worldwide.

22 Baskin, "Northern Exposure," 2008.

23 Peterson, G., "New Statute for Canadian Oil Sands," *Geotimes* 48, no. 3 (2003): 7.

24 Approximately 1.0 to 1.25 gigajoules of natural gas are needed per barrel of bitumen extracted. A barrel of oil equivalent is about 6.117 gigajoules, so this produces about five or six times as much energy as is consumed. From "FAQ: Oil Sands," 2008. http://environment.gov.ab.ca/info/faqs/faq5-oil_sands.asp.

25 Freight statistics are from Department of Transportation, Table 2-1: "Weight of Shipments by Transportation Mode: 2002, 2007, and 2035." http://ops.fhwa. dot.gov/freight/freight_analysis/nat_freight_stats/docs/08factsfigures/tabl e2_1.htm. Accessed 2 September 2009.

26 Vardi, Nathan, "Crude Awakening," *Forbes,* 27 March 2006.

27 Wikipedia, as of 2006–2007 (the most recent data available). In metric, $420km^2$ have been affected.

28 "FAQ: Oil Sands," 2008. http://environment.gov.ab.ca/info/faqs/faq5-oil_sands. asp.

29 *Ibid.*

30 "The Most Destructive Project on Earth: Alberta's Tar Sands," Celsias website. www.celsias.com/article/the-most-destructive-project-on-earth-albertas-tar/. (Note that this is a discussion of and reference to the report by Hatch and Price listed below.)

31 Hatch, Christopher, and Matt Price, "The Most Destructive Project on Earth" (New York City: Environmental Defense, 2008). Polycyclic aromatic hydrocarbons are a group of ring-compounds, February 2008. http://www.environmental defence.ca/reports/tarsands.htm.

32 *Ibid.*

33 *Ibid.*

34 Birger, Jon, "Oil from a Stone," *Fortune,* 1 November 2007. http://fortunemagazineng.com/enzineport/content.asp?contenttype=maincontents.

35 Sengupta, Somini, "Indians Hit the Road Amid Elephants," *New Delhi Journal,* 11 January 2008.

36 "China," in *Encyclopedia Britannica.* www.britannica.com/eb/article-257894. Accessed 11 March 2008.

37 "Investing in China's Booming Automobile Sector: Japanese Cars a No Go," Seeking Alpha website, posted 19 March 2007. Seeking Alpha website available at http://seekingalpha.com/article/29954-investing-in-china-s-booming-automobile-sector-japanese-cars-a-no-go.

38 © 2008 Associated Press. The information contained in the AP news report may not be published, broadcast, rewritten, or otherwise distributed without the prior written authority of the Associated Press. Active hyperlinks have been inserted by AOL (captured 3 March 2008, 6:11 a.m. EST).

39 Zhao, Jimin, "Can the Environment Survive China's Craze for Automobiles?" School of Natural Resources and Environment, University of Michigan, (Submitted to Transportation Research Part D: Transport and Environment.) www.cebc.org.br/sites/500/522/00000349.pdf.

40 EIA Table 5.3, "Average Retail Price of Electricity to Ultimate Customers: Total by End-Use Sector, 1995 through May 2009," gives the value in the text. (Sources: Energy Information Administration, EIA-826, "Monthly Electric Sales and Revenue Report with State Distributions Report: 2006–2008"; and EIA-861, "Annual Electric Power Industry Report: 1992 - 2005.")

According to the U.S. Department of Energy, the average delivered cost for coal, petroleum, and natural gas used for electricity generation increased between 2004 and 2005. The average cost of natural gas to electricity generators increased from the previous record high of $5.96 per million BTU (MMBTU), established in 2004, to a new record level of $8.21 per MMBTU in 2005. For the third year in a row, natural gas costs experienced a double-digit percentage increase, 37.8% from 2004 to 2005. As a result, the cost of natural gas for electricity generation in 2005 was 130.6% higher than in 2002.

The average delivered cost of coal increased 13.2% for the year, in part due to increases in coal mining operations and the cost of electricity and diesel fuels for that mining. The average delivered cost for all fossil fuels used for electricity generation (coal, petroleum, and natural gas combined) in 2005 was 114.5% higher than in 2002 (reported at www.eia.doe.gov/cneaf/electricity/epa/epa_sum.html on 21 March 2007).

41 "Aramco, Dow Chemical sign huge deal," *Wall Street Journal,* 15 May 2007, International Edition, 30.

42 Campoy, Ana, and Leslie Eaton, "Chemical Prices Jump, Fueling Fear of Inflation," *Wall Street Journal,* 29 May 2008, A1.

43 Graham, Sarah, "Environmental Effects of *Exxon Valdez* Spill Still Being Felt," *Scientific American* 292 (2003):12–19.

44 The history of litigation over the *Exxon Valdez* oil spill is interesting because of its length and the failure for it to provide much of a payment to those who suffered from the spill. http://en.wikipedia.org/wiki/Exxon_Valdez_oil_spill.

Chapter 2

1 In this image made from video provided by KHOU-TV on May 7, 2008, a large tank, center, falls into a sinkhole near Daisetta, Texas (AP Photo/KHOU-TV, Bobby Bracken).

2 Krauss, C., "Natural Gas Has Utah Driving Cheaply," *New York Times,* August 30, 2008.

3 "The Pickens Plan" at www.pickensplan.com/theplan/. Accessed 21 September 2008.

4 Krauss, "Natural Gas Has Utah Driving Cheaply," 2008.

5 Annual use of natural gas in the United States in 2007 totaled 23,055,596 million cubic feet. Ray Boswell, Ph.D., Manager, Methane Hydrate R&D Programs, U.S. Department of Energy—National Energy Technology Laboratory. Dr. Boswell

provided much of the basic information about natural gas reserves and rate of use, and discussed with me at length how these quantities are estimated.

6 Mouawad, Jad, "Estimate Places Natural Gas Reserves 35% Higher," *New York Times,* 18 June 2009.

7 U.S. EIA, "Worldwide Look at Reserves and Production," *Oil & Gas Journal* 106, no. 48 (22 December 2008): 22–23. www.eia.doe.gov/oiaf/ieo/nat_gas.html.

8 Geological Survey of Canada. http://gsc.nrcan.gc.ca/gashydrates/canada/index_e.php. Modified 12 December 2007.

9 Dillon, Dr. William, and Dr. Keith Kvenvolden, *Gas (Methane) Hydrates—A New Frontier* (Washington, D.C.: USGS, September 1992. http://marine.usgs.gov/fact-sheets/gas-hydrates/title.html.

10 Botkin, D. B., and E. A. Keller, *Environmental Science: Earth as a Living Planet, 7th edition* (New York: John Wiley & Sons: 2009). See Chapter 18, on fossil fuel energy.

11 Gold, R., "Gas Producers Rush to Pennsylvania: Promising Results for Wells There Spur Investment," *Wall Street Journal* 2 April 2008, A2.

12 Krauss, C., "Drilling Boom Revives Hopes for Natural Gas," *New York Times,* August 24, 2008.

13 *Ibid.*

14 Casselman, B., "Texas Sinkhole Puts Spotlight on Oil, Gas Drilling," *Wall Street Journal,* May 19, 2008, A3.

Chapter 3

1 Brady photograph plate 113; Washington, D.C. streetlamp, 1865; Library of Congress. Also National Park Service Publication, "Gas Lighting in America: A Guide for Historic Preservation." www.nps.gov/history/history/online_books/hcrs/myers/plate12.htm. Accessed 25 March 2008.

2 FutureGen Alliance website, www.futuregenalliance.org/news/response_to_doe_rfi_030308.stm. Updated January 2008. Accessed 25 March 2008.

3 Sherman, Mimi, "A Look at Nineteenth-Century Lighting: Lighting Devices from the Merchant's House Museum," APT Bulletin of the Association for Preservation Technology International Lighting Historic House Museums 1 (2000): 37–43.

4 "Coal Gasification," FutureGen Alliance website, www.futuregenalliance.org/technology/coal.stm. Accessed 25 March 2008.

5 Anonymous (2010). "Exelon joined FutureGen in January." Reuters News Service. (New York, Reuters News Service.)

6 Skinner, B., S. Porter, and D. B. Botkin, *The Blue Planet* (New York: John Wiley & Sons, 1999). Many popular references get the time when coal formed wrong.

7 Some of the basic facts about coal come from the American Coal Foundation; see www.teachcoal.org/aboutcoal/articles/faqs.html. Accessed 19 March 2008. Other facts come from DOE EIA and the World Coal Institute.

8 Coal mining data in the United States is for 2006. Source: Energy Information Administration, *Quarterly Coal Report, October–December 2008,* (Washington, D.C.; April 2009). www.eia.doe.gov/cneaf/coal/page/special/fig1.html.

9 World Coal Institute, *Coal Facts 2007.* www.worldcoal.org/pages/content/index.asp?PageID=188.

10 Coal mining data in the United States is for 2006. Source: Energy Information Administration, www.eia.doe.gov/fuelcoal.html. Accessed 19 March 2008.

11 Coal use in the United States is from the DOE EIA *Annual Coal Report.* www.eia.doe.gov/cneaf/coal/page/acr/acr_sum.html#fes1. Accessed 19 March 2008.

12 The data I have used in my calculations, from the BP statistical data, gives an estimate of 300 years for coal, but the World Coal Institute states that coal will last 150 years. (See World Coal Institute, *Coal Facts 2007,* at www.worldcoal.org/pages/content/index.asp?PageID=188.) This range of estimates is typical and to be expected with data as complex and difficult to obtain as the total reserves of coal. In fact, statements that lack such ranges in this context are likely to be less scientifically accurate and reliable and less trustworthy.

13 Hooke, Roger Leb, "Spatial Distribution of Human Geomorphic Activity in the United States: Comparison with Rivers," *Earth Surface Processes and Landforms* 24 (1999): 687–692. I use his worldwide figures of rivers moving 14GT per year (not including 10GT per year from agriculture) and the amount that people move worldwide, at 35GT. Using this ratio, 14/35 = 40% gives a value of 3.04GT per year in relation to the 7.6GT per year moved in the U.S.

14 Hooke, "Spatial Distribution of Human Geomorphic Activity," 1999.

15 "Coal," *Encyclopedia of Appalachia* (Knoxville, Tenn.: University of Tennessee Press, 2008).

16 Appalachian Voices website, www.appvoices.org/index.php?/frontporch/blogposts/environmental_groups_ask_un_to_oppose_appalachian_coal_mining_practices/.

17 Caudill, Harry M. *Night Comes to the Cumberlands: A Biography of a Depressed Area* (Boston: Little, Brown and Company; 1963).

18 *Ibid.,* p. 306–207.

19 *Ibid.,* p. 318.

20 Diehl, Peter, "The Inez Coal Tailings Dam Failure (Kentucky, USA)," WISE Uranium Project, part of World Information Service on Energy, 2008.

21 Bingham, Barry, Jr., "Mining Is Turning Eastern Kentucky into a Despicable Latrine," *Louisville Courier Journal,* November 9, 2005.

22 Rahn, P. H., *Engineering Geology: An Environmental Approach* (New York: Elsevier, 1986).

23 Botkin, D. B., and E. A. Keller, *Environmental Science: Earth as a Living Planet, 7th Edition* (New York: John Wiley & Sons, 2009).

24 www.undergroundminers.com/laurelrun.html.

25 www.offroaders.com/album/centralia/other-mine-fires.htm.

26 Netherlands Earth Observation, *Environmental Monitoring of Coal Fires in North China Project Identification Mission Report*, October 1993. http://apex. neonet.nl/browse/www.neonet.nl/Document/XHCFRJGIVMUWUYOTOVMW SXKLG.html.

27 "How China's Scramble for 'Black Gold' is Causing a Green Disaster," *Daily Telegraph*, 01 Feb 2002.

28 Revkin, Andrew C., "Sunken Fires Menace Land and Climate," *New York Times*, 15 January 2002. http://query.nytimes.com/gst/fullpage.html?res=9902 E2DF1538F936A25752C0A9649C 8B63.

29 *Ibid.*

30 www.nrdc.org/globalWarming/coal/contents.asp.

31 "Clean Air, Dirty Coal," Sierra Club website, www.sierraclub.org/cleanair/ factsheets/power.asp. Accessed 23 March 2008.

32 NRDC, *Return Carbon to the Ground: Reducing Global Warming Pollution and Enhancing Oil Recovery*, 2006. www.nrdc.org/globalwarming/solutions.

33 For additional information, see "Clean Air, Dirty Coal," Sierra Club website, www.sierraclub.org/cleanair/factsheets/power.asp.

34 Grand Canyon Trust, www.grandcanyontrust.org/programs/air/mohave.php. Accessed 24 March 2008.

35 Department of Interior Office of Surface Mining website, www.wrcc.osmre.gov/ BlkMsaQ_A/BMFAQ.htm. Accessed 19 March 2008.

36 Black Mesa Indigenous Support. www.blackmesais.org/struggle_continues05. htm. This organization describes itself as "First Nations, First Resistance— Support the Struggle for Survival at Big Mountain, Black Mesa, Ariz."

    "On behalf of their peoples, their ancestral lands, and future generations, more than 350 Dineh residents of Black Mesa continue their staunch resistance to the efforts of the U.S. Government—acting in the interests of the Peabody Coal Company—to relocate the Dineh and destroy their homelands."

37 Southern California Edison website, www.sce.com/PowerandEnvironment/ PowerGeneration/MohaveGenerationStation/. Accessed 24 March 2008.

38 Grand Canyon Trust, www.grandcanyontrust.org/programs/air/mohave.php. Accessed 24 March 2008.

39 Frey, Steve, *Nevada Visibility FIP for Nevada* (Washington, D.C.: EPA, 2001).

40 Grand Canyon Trust, www.grandcanyontrust.org/programs/air/mohave.php. Accessed 24 March 2008.

41 "Mohave Power Plant Set to Close," United Press International. "In a filing Thursday with the California Public Utilities Commission, Edison said it wanted to continue negotiations to keep the power plant open, to add pollution controls that are expected to cost $1 billion, but close for at least a few months." www.physorg.com/news9480.html, 31 December 2005.

    Also see www.physorg.com/news9480.html. © 2005 United Press International.

42 Bureau, Kathy Helms Diné, "Mohave Power Plant Looking at Restarting," *The Gallup Independent*, 9 July 2007.

43 www.blackmesais.org/bigmtbackground.html.

44 Office of Surface Mining, www.osmre.gov/amlgrant04.htm. Accessed 20 March 2008.

45 Office of Surface Mining, www.osmre.gov/reggrants98.htm.

46 Smith, R. (2009). "U.S. Foresees a Thinner Cushion of Coal," *Wall Street Journal*. New York, Dow Jones.

47 EIA Report 058 (2008). September, 2008.

48 Schlissel, D. Allison Smith, and Rachel Wilson, *Coal-Fired Plant Construction Costs*, Synapse Energy Economics, Inc. 2008.

49 DOE EIA, "Net Generation by Energy Source: Total (All Sectors)," Electric Power Monthly with data for October 2009; Report Released: January 15, 2010. Coal produced about 2 billion megawatt-hours a year, each year since 2000. In 2009, coal produced 1.99 billion megawatt hours. http://www.eia.doe.gov/cneaf/electricity/epm/table1_1.html

50 NRDC, www.nrdc.org/coal/19c.asp. Accessed 23 March 2008.

51 NRDC, (2006).

52 Anonymous, "Mountaintop Advocates Open New Front in Fight Against Coal—Challenge Billion-Dollar Government Giveaways for Not Considering Cost to the Mountains." 3 March 2008. www.ilovemountains.org/all/371.

53 Katzer, James, et al. *The Future of Coal: Options for a Carbon-Constrained World* (Boston: Massachusetts Institute of Technology, 2007).

54 Wald, Matthew L., "Two Utilities Are Leaving Clean Coal Initiative," *New York Times*, 26 June 2009. "Two of the nation's biggest coal-burning utilities said Thursday that they were withdrawing from a $2.4 billion project to demonstrate carbon capture and storage, and would instead pursue their own work in the field."

55 Madrigal, Alexis, "Back to the FutureGen: 'Clean' Coal Plant Gets New Backing," *Wired*, 12 June 2009. www.wired.com/wiredscience/2009/06/futuregen/.

56 Canine, Craig, "How to Clean Coal," *ONEARTH* NDRC online Magazine, Fall 2005. http://www.nrdc.org/OnEarth/05fal/coal1.asp.

57 Information about the Greenpoint coal liquification comes from Shogren, Elizabeth, "Turning Dirty Coal into Clean Energy," National Public Radio, 25 March 2008.

58 *Ibid.*

Chapter 4

1 Photo of Edwards Dam is from the Government of Maine, 7 March 2007. www.maine.gov/spo/sp/edwards/progress.php.

2 American Rivers, *Restoring Rivers*, 2007. (Major upcoming dam removals in the Pacific Northwest). www.water.ca.gov/fishpassage/docs/dams/dams.pdf.

3 *The World Bank & The World Commission on Dams Report Q&A*, World Bank Publications, 2001. http://web.worldbank.org/WBSITE/EXTERNAL/TOPICS/EXTWRM/0, contentMDK:2043 8903~pagePK:148956~piPK:216618~theSite PK:337240,00.html. Accessed 5 March 2007.

4 Botkin, D. B., 1999, "When Should a Dam be Breached?" *Los Angeles Times*, Sunday, August 22, 1999.

5 U.S. DOE EIA website, http://www.eia.doe.gov/cneaf/solar.renewables/page/hydroelec/hydroelec.html. Accessed 12 February 2010. This source states that the total generation was 206,148 thousand megawatt-hours in 2009.

6 Botkin, D. B., and E. A. Keller, *Environmental Sciences: The Earth as a Living Planet* (New York: John Wiley & Sons, 2009).

7 World Commission on Dams, *Dams and Development: a New Framework for Decision Making*.

8 National Renewable Energy Laboratory.

9 Wisconsin Valley Improvement Company website, www.wvic.com/hydro-facts.htm. Accessed 8 February 2008.

10 The Pacific Coast Federation of Fishermen's Associations website. www.pcffa.org/dams.htm. Accessed 5 February 2008.

11 Barry Goldwater said this when asked by *Vanity Fair* to name his "greatest political regret."

12 "The potential for hydroelectric power projects on the Nam Theun was first identified in the mid-1970s and was the subject of detailed studies during the following decades. It was not until the early 1990s that the Nam Theun 2 hydroelectric project (NT2 or the Project) was specifically recognized by the Government of the Lao PDR as a key project for the economic and social development of the Lao nation." http://www.namtheun2.com/. Accessed February 8, 2008.

13 World Bank, *The Nam Theun 2 Hydroelectric Project (NT2): An Overview and Update* (Washington, D.C.; World Bank, 2006).

14 *The World Bank & The World Commission on Dams Report Q&A.*

15 The World Bank estimates that revenues will rise to $110 million from 2020 to 2034.

16 According to the *CIA World Fact Book*, Laos electricity consumption is 1.715 billion kilowatt-hours (2005); production is 1.193 billion kilowatt-hours (2005).

17 Scudder, T., *The Future of Large Dams: Dealing with Social, Environmental, Institutional, and Political Costs* (London: Earthscan, 2006).

18 For additional reading about environmental and social effects of large dams, see Leslie, J., *Deep Water: The Epic Struggle Over Dams, Displaced People, and the Environment* (New York: Farrar Straus Giroux, 2005).

19 http://internationalrivers.org/en/follow-money/world-bank/nam-theun-2-investigation-exposes-project-failings. Accessed 10 February 2008.

20 Lawrence, S., "Doing Dams Wrong: World Bank's Model Project Leaves Lao Villagers in the Lurch," *World Rivers Review* (2007): 10–15.

21 Botkin and Keller, *Environmental Sciences*, 2009; see Chapter 21.

22 Scudder, T., 2006. *The Future of Large Dams: Dealing with Social, Environmental, Institutional, and Political Costs* (London: Earthscan, 2006.) See Box 9.1, "Key Message," page 283.

23 Timmons, H., "Energy from the Restless Sea," *New York Times*, 3 August 2006.

24 Based on information from the U.S. DOE and EPRI.

25 Values are for 2005, the most recent data, from Energy Information Administration, *International Energy Annual 2005,* table posted 2 October 2007.

26 Energy Information Administration, *International Energy Annual 2004* (May–July 2006), www.eia.doe.gov/international. Lists the information for the U.S. in Table 1.3 and for other countries in Table 1.8. Values are for 2005, the most recent data.

27 International Hydropower Association, *Hydropower and the World's Energy Future: The Role of Hydropower in Bringing Clean, Renewable Energy to the World,* November 2000. This report cites information from *Hydropower & Dams, World Atlas and Industry Guide,* 2000. These are the specifics I used for the text's discussion: The world's total technical feasible hydro potential is estimated at 14,370 billion kilowatt-hours per year, of which about 8,082 billion kilowatt-hours per year is currently considered economically feasible for development. About 700 million kilowatt-hours (or about 2,600 billion kilowatt-hours per year) is already in operation, with a further 108 million kilowatt-hours under construction. Most of the remaining potential is in Africa, Asia, and Latin America. Translated into simple English, water power provides about 2% of the world's total energy, and about one-third (36%) of the world's possible water-power sites have been developed.

28 *Ibid.*

29 *Ibid.*

30 *Restoring Rivers,* 2007. www.water.ca.gov/fishpassage/docs/dams/dams.pdf.

Chapter 5

1 Wald, Matthew L., "Foes of Indian Point Begin Legal Battle," *New York Times,* 11 March 2008.

2 "Nuclear power's most effective spokesman may be Patrick Moore, a founder and former member of the environmental group Greenpeace, who has been hired by the nuclear industry to promote the technology." Applebome, P., "The Power Grid Game: Choose a Catastrophe," *New York Times,* 9 December 2007.

3 Moor, Patrick, "Going Nuclear: A Green Makes the Case," *Washington Post,* 16 April 2006, B01.

4 Franz J. Dahlkamp email. www.independent.co.uk/opinion/commentators/hugh-montefiore-we-need-nuclear-power-to-save-the-planet-from-looming-catastrophe-544571.html.

5 Dr. Franz J. Dahlkamp, the world's leading expert on uranium ore and the author of a five-volume work on the subject, responded to my inquiry, in which I asked for the best sources on this topic. He recommended two sources in his email, dated 14 July 2008: AEA, Analysis of Uranium Supply to 2050 (Vienna: AEA, 2001); and OECD-NEA & IAEA, Uranium 2005: Resources, Production and Demand (Paris: OECD, 2005).

6 Estimates of the Years That Nuclear Power Plant Fuel Will Last are based on the IAEA estimates of Uranium Ore Reserves.

7 According to the International Atomic Energy Agency, at www.iaea.org/inis/aws/
   fnss/auxiliary/iaea.html. Accessed 14 July 2009.

   "For three decades several countries had important fast breeder reactor development programs. Fast test reactors (Rapsodie [France], KNK-II [Germany],
   FBTR [India], JOYO [Japan], DFR [UK], BR-10, BOR-60 [Russia], EBR-II,
   Fermi, FFTF [USA]) were operating in several countries, with commercial size
   prototypes (Phènix, Superphènix [France], SNR-300 [Germany], MONJU
   [Japan], PFR [UK], BN-350 [Kazakhstan], BN-600 [Russia]) just under construction or coming on line. However, from the 1980s onward, and mostly for econom-
   ical and political reasons, fast reactor development in general began to decline.
   By 1994, in the USA, the Clinch River Breeder Reactor (CRBR) had been can-
   celed, and the two fast reactor test facilities, FFTF and EBR-II, had been shut
   down—with EBR-II permanently and FFTF, until recently, in standby condition,
   but now also facing permanent closure. Thus, in the U.S., effort essentially disappeared for fast breeder reactor development. Similarly, programs in other nations
   were terminated or substantially reduced. In France, Superphènix was shut down
   at the end of 1998; SNR-300 in Germany was completed but not taken into operation, and KNK-II was permanently shut down in 1991 (after 17 years of operation) and is scheduled to be dismantled by 2004. In the UK, PFR was shut down
   in 1994, and in Kazakhstan, BN-350 was shut down in 1998."

8 According to a Wikipedia article, which I have not verified independently, "As of
   2003 one indigenous FBR [breeder reactor] was planned for India, and another
   for China. Both were to use Soviet technology. Meanwhile, South Korea was said
   to be designing a standardized modular breeder reactor for export. The FBR program of India includes the concept of using fertile thorium-232 to breed fissile
   uranium-233. Also, a Russian breeder reactor was said to be still operational in
   Zarechny. And on February 16, 2006, the U.S., France, and Japan signed an
   'arrangement' to research and develop sodium-cooled fast reactors in support of
   the Global Nuclear Energy Partnership."

9 EIA, *International Energy Annual 2003,* July 2005.

10 EIA, *System for the Analysis of Global Energy Markets,* 2006.

11 Nuclear Energy Information Service, "Nuclear Power Has Cost This Country
   over \$492,000,000,000." www.neis.org/literature/Brochures/npfacts.htm.
   Accessed 25 April 2008.

12 Smith, Rebecca, "U.S. Chooses Four Utilities to Revive Nuclear Industry," *Wall
   Street Journal,* 17 June 2009, A1.

13 World Health Organization, www.who.int/mediacentre/news/releases/2005/pr38/
   en/index1.html. Accessed 1 June 2009.

   "The total number of deaths already attributable to Chernobyl or expected in the
   future over the lifetime of emergency workers and local residents in the most
   contaminated areas is estimated to be about 4,000. This includes some 50 emergency workers who died of acute radiation syndrome and nine children who died
   of thyroid cancer, and an estimated total of 3,940 deaths from radiation-induced
   cancer and leukemia among the 200,000 emergency workers from 1986–1987,
   116,000 evacuees, and 270,000 residents of the most contaminated areas (total
   about 600,000). These three major cohorts were subjected to higher doses of
   radiation amongst all the people exposed to Chernobyl radiation."

14 Patel, Julie, "FP&L Might Be Fined over Nuclear Plant Security: Security Workers Dozed, Regulators Say," *Sun-Sentinel,* 11 April 2008. www.sun-sentinel.com/business/sfl-flzfpl0411sbapr11,0,2008712.story South Florida Sun-Sentinel.com.

15 World Nuclear Organization, "Waste Management in the Nuclear Fuel Cycle," www.world-nuclear.org/info/inf04.html.

16 IAEA International Atomic Energy Agency. "South Africa Hosts Global Workshop on Radioactive Waste: Looking to Forge Common Approach for Management and Disposal Policies Staff Report." June 25, 2007. IAEA News Center. The report states that there are 200,000 metric tons of wastes, which I have converted to British tons (2,000 pounds each). See www.iaea.org/NewsCenter/News/2007/saradwaste.html.

Note that this information comes from the International Atomic Energy Agency, which describes itself as being "set up as the world's 'Atoms for Peace' organization in 1957" within the United Nations.

17 World Nuclear Association, "Waste Management in the Nuclear Fuel Cycle." www.world-nuclear.org/info/inf04.html. Accessed 16 May 2008. The World Nuclear Association gives a much higher figure for total nuclear wastes in storage: 270 metric tons.

18 OECN, International Conference on Management of Spent Fuel from Nuclear Power Reactors, Vienna, Austria, 19–23 June 2006.

19 World Nuclear Organization, 2007.

20 Richardson, Ingela, "Filthy Lucre—Nuclear Waste Costs Lives," Coalition Against Nuclear Power, 10 September 2007. Once again, the report is in metric units, so it gives 2 million metric tons.

21 World Nuclear Organization, 2007.

22 Alliance for Nuclear Responsibility, 2008. For the mission statement, see http://a4nr.org/elements/elements/mission. For information on decay rates, see http://a4nr.org/library/lowlevel/nirs.lowlevelradioactivewaste.

23 Kestenbaum, David, "EPA Expected to Issue Million-Year-Long Regulation," National Public Radio, 24 November 2006. http://mustv.com/templates/story/story.php?storyId=6525491.

24 *CBS News* story on Yucca Mountain (July 25, 2004). www.cbsnews.com/stories/2003/10/23/60minutes/main579696.shtml.

25 Halstead, Bob, Dave Ballard, Hank Collins, and Marvin Resnikoff, *State of Nevada Perspective on the U.S. Department of Energy Yucca Mountain Transportation Program* (Phoenix, Ariz.: Waste Management, 2008).

26 *CBS News,* Yucca Mountain, 2004.

27 Hughes, Siobhan, "U.S. House Votes Against Eliminating Yucca Mountain Funding," *Wall Street Journal* online, 17 July 2009. http://online.wsj.com/article/BT-CO-20090717-711221.html.

28 Halstead, et al., *State of Nevada Perspective,* 2008.

29 Matthew L. Wald, "Obama Acts to Ease Way to Construct Reactors," *New York Times,* 29 January 2010.

30 Godoy, Julio, "Environment: France's Nuclear Waste Heads to Russia," Inter Press Service, Dec 17, 2005. The Inter Press Service, headquartered in Rome, calls itself "a communication institution with a global news agency at its core," raising "the voices of the south and civil society." See http://ipsnews.net/news.asp?idnews=31466. Accessed 16 May 2008. "According to the study 'La France nucléaire,' published in 2002 by the World Information Service on Energy (WISE), each year the French nuclear station Eurodif, situated on the banks of the Rhone River, 700 km south of the French capital, produces 15,000 tonnes of depleted uranium. "Most of that waste is of no further use, and is simply stored at the nuclear plant. Today there are an estimated 200,000 tonnes of this nuclear material being warehoused there."

31 World Nuclear Association, "Radioactive Wastes," March 2001. www.world-nuclear.org/info/inf60.html?terms=vitrified+wastes. "High-level Waste (HLW) contains the fission products and transuranic elements generated in the reactor core which are highly radioactive and hot. High-level waste accounts for over 95% of the total radioactivity produced though the actual amount of material is low, 25–30 tonnes of spent fuel, or three cubic metres per year of vitrified waste for a typical large nuclear reactor (1000 MWe, light water type)."

32 Greenpeace International, "Illegal French Nuclear Waste Dump Must Be Removed and Decontaminated." www.greenpeace.org/international/press/releases/illegal-french-nuclear-waste-d. Accessed 29 May 2006.

33 Greenpeace International, "Radioactive Waste Leaking into Champagne Water Supply Levels Set to Rise, Warns Greenpeace." www.greenpeace.org/international/press/releases/illegal-french-nuclear-wasted.

34 The actual quote from Greenpeace is, "On April 22, 2005, ANDRA informed the French nuclear safety authority DGSNR that the wall of a storage cell fissured while concrete was added on the last layer of wastes stored in the CSA disposal site. The origin of the fissure was a 'water corner' phenomenon resulting from the hydrostatic pressure of a water column formed with the infiltration and which could lead to the breaking of the wall. The DGSNR have admitted that this 'water corner' phenomenon was under-evaluated during the conception of some cells. The nuclear safety Authority demanded that all these cells be from now on conceived to resist the most severe 'water corner' phenomenon. Regarding the cells already built, the setting of a surrounding waterproof joint at each concrete layer will prevent this phenomenon from happening. This event revealed a flaw in the conception of the storage cells of the site." Full copy available in French and English at www.greenpeace.fr.org and www.stop-plutonium.org.

35 World Health Organization, www.who.int/mediacentre/news/releases/2005/pr38/en/index1.html. Accessed 1 June 2009.

36 See www.findingdulcinea.com/news/on-this-day/March-April-08/On-this-Day-Chernobyl-Nuclear-Power-Plant-Melts-Down.html?gclid=CN7mu6Hiq5MCFQKaFQodg3nu3g.

37 For updates on Chernobyl, see the International Atomic Energy Agency (IAEA) website; such as "The Enduring Lessons of Chernobyl by IAEA Director General Dr. Mohamed ElBaradei." 6 September 2005. www.iaea.org/NewsCenter/Statements/2005/ebsp2005n008.html.

Section II

1 DOE EIA. Table 1. U.S. Energy Consumption by Energy Source, 2003–2007. www.eia.doe.gov/cneaf/alternate/page/renew_energy_consump/table1.html.

2 DOE EIA. www.eia.doe.gov/cneaf/solar.renewables/page/prelim_trends/rea_prereport.html.

Chapter 6

1 In a west Texas field, wind turbines generate electricity (© iStockphoto.com/ chsfoto).

2 Dodge, Darrell M., *Illustrated History of Wind Power Development*, Chapter 1. http://www.telosnet.com/wind/.

3 Wind Power History, "Wind Power's Beginnings (1000 B.C.–1300 A.D.)." www.telosnet.com/wind/early.html.

4 SkySails website, www.skysails.info/index.php?id=64&L=1&tx_ttnews[tt_news]=98&tx_ttnews[backPid]=6&cHash=c1a209e350. Accessed 31 March 2008.

5 Information about the SkySails project comes from the company's website and press release; www.skysails.info/index.php?id=64&L=1&tx_ttnews[tt_news]=98&tx_ttnews[backP id]=6&cHash=c1a209e350 and www.skysails.info/index.php?id=64&L=1&tx_ttnews[tt_news]=104&tx_ttnews[back Pid]=6&cHash=db100 ad2b6; and Herron, James, "Wind Makes a Return to Power the Beluga on 'Greener' Journey," *Wall Street Journal,* 21 January 2008.

6 The United States uses 1.42 thousand billion kilowatt-hours a years (1.42 × 1,012 kilowatt-hours).

7 FPL, Mountaineer Wind Energy Center, Florida Power and Light, 2008. www.fplenergy.com/portfolio/pdf/mountaineer.pdf.

8 DOE Photograph. http://www.doedigitalarchive.doe.gov/SearchImage.cfm?page=search.

9 Gipe, Paul, "One Million Megawatts of Wind Capacity for the USA: A Target Worthy of a Great Nation," www.wind-works.org/LargeTurbines/OneMillion-MegawattsofWindCapacity.html. Accessed 23 January 2008. For further information, see Gipe's revised book, *Wind Energy Basics, Second Edition: A Guide to Home- and Community-Scale Wind Energy Systems* (Chelsea Green Publishing 2009).

10 Infoplease, "Top 50 Cities in the U.S. by Population and Rank." www.infoplease.com/ipa/A0763098.html. Accessed 17 June 2008. (© 2000–2007 Pearson Education).

11 Texas Energy Conservation Office, "2008 Texas Wind Energy." www.seco.cpa.state.tx.us/re_wind.htm. Accessed 30 April 2008.

12 The calculation gives 1,035,294 of these turbines needed.

13 Tegen, S., M. Goldberg, and M. Milligan, "User-Friendly Tool to Calculate Economic Impacts from Coal, Natural Gas, and Wind: The Expanded Jobs and Economic Development Impact Model (JEDI II)" (paper presented at WIND-POWER 2006, Pittsburgh, Penn., June 2006). "For example, a new coal plant in

South Dakota (Big Stone II) is priced at approximately $1,900 per kilowatt, whereas a new coal plant in Colorado (Comanche III) is estimated to cost less than $1,500 per kilowatt.

14 Florida Power and Light, "Economics of Wind Energy." www.fplenergy.com/portfolio/wind/economics.shtml. Accessed 25 April 2008.

15 DOE EIA Table ES1. Summary Statistics for the United States, 1994 through 2005 in computer file DOE Statistics Copy of epaxlfilees1.xls.

16 The total capacity for the production of electrical energy in the United States is 978,020MW. Of this, 754,989MW comes from fossil fuel power plants.

17 Stiglitz, Joseph E., "The True Costs of the Iraq War," Project Syndicate, 2006. www.project-syndicate.org or www0.gsb.columbia.edu/ipd/pub/JES_paper.pdf. Joseph E. Stiglitz, a Nobel laureate in economics, is a professor of economics at Columbia University and was the chairman of the Council of Economic Advisers to President Clinton, as well as the chief economist and senior vice president at the World Bank.

18 World Wind Energy Association website stated that the Worldwide capacity in 2007 was 93,8 GW with 19,7 GW added in 2007. http://www.wwindea.org/home/index.php Accessed 21 February 2008.

19 *Ibid.*

20 www.wwindea.org/home/index.php.

21 Anonymous, "Enron Acquires Zond, a Major Wind-Power Company," *New York Times*, 7 January 1997.

22 www.awea.org/projects/.

23 Between 1999 and 2008, wind-power generation capacity in the United States increased sixfold, from 2,500 million watts to more than 16,000 million watts, and more than 3,600 million watts are under construction. Only 11 states had large-scale wind-power installations in 1999; today more than 30 states have them.

24 Some European nations are committed to wind energy, including Britain, Canada, Denmark, Germany, Italy, Japan, the Netherlands, Norway, Spain, and Sweden. Most impressive is Spain's use of wind energy. A major milestone was reached for world wind energy on April 19, 2008, when wind produced 10,879MW in Spain, more than one-third of the nation's total electricity production; nuclear power was second.

25 Spanish Wind Energy Organization. See www.aeeolica.org/ and http://actualidad.terra.es/nacional/articulo/record_absoluto_produccion_eolica_porcentaje_2411417.htm.

26 Landler, M., "Sweden Turns to a Promising Power Source, With Flaws," *New York Times*, 23 November 2007.

27 Childress, S., "Electrifying a Nation, Mr. Kamkwamba's Creation Spurs Hope in Malawi; Entrepreneurs Pay Heed," *Wall Street Journal*, 12 December 2007, A1.

28 William Kamkwamba's blog, http://williamkamkwamba.typepad.com/williamkamkwamba/2007/06/welcome_to_my_b.html29. Also see Stimmel, R., *Small Wind Turbine Global Market Study* (American Wind Energy Association, 2007). www.awea.org/smallwind/documents/AWEASmallWindMarketStudy2007.pdf.

30 *Ibid.*

31 Galbraith, Kate, "North Carolina: Effort to Ban Wind Turbines," *New York Times*, 8 August 2009.

32 Cassidy, P., "Wind Farm Generates More Than 40,000 Comments," *Cape Cod Times*, 23 April 2008.

33 Reuters, "Wind Farm Clears Hurdle," 15 January 2008.

34 Galbraith, Kate, "Texas Is More Hospitable Than Mass. to Wind Farms Economy, Culture Fueling a Boom," *The Boston Globe*, 25 September 2006.

35 Cassidy, P., "Floating Wind Farm Plan Dealt Blow," *Cape Cod Times*, 19 April 2008.

36 Zezima, K. "Interior Secreatary Sees Little Hope for Consensus on Wind Farm," *New York Times*, 2008.

37 See Senator Lamar Alexander, "Blueprint for 100 New Nuclear Power Plants in 20 Years: How Nuclear Power Can Provide Enough Clean, Cheap, Reliable, American Energy to Create Jobs, Clean the Air, and Solve Global Warming," pre-publication draft report, 29 July 2009.

38 Gipe, Paul, www.wind-works.org/articles/NRELBirdReport04.html.

39 Nijhuis, Michelle, "Selling the Wind," Audubon Society, 2008. http://audubonmagazine.org/features0609/energy.html. Accessed 28 April 2008.

40 *Ibid.*

Chapter 7

1 Panasonic World Solar Challenge official final results.

2 Energy Information Administration, www.eia.doe.gov/cneaf/solar.renewables/page/solarphotv/solarpv.html. Accessed 2 June 2009.

3 Photovoltaic thin films use very small amounts of certain rare metal compounds, including cadmium telluride (a compound of cadmium and tellurium), CIGS, and microcrystalline silicon. Arguably, the most successful of these to date is cadmium telluride because it is less expensive to manufacture than other photovoltaics (First Solar Corporation, www.firstsolar.com/material_sourcing.php, accessed 2 June 2009). Both cadmium and tellurium are obtained as byproducts in the mining of other metals. Although cadmium is toxic to many life forms, the amount used is small and embedded in glass. According to a study done at Brookhaven National Laboratory, this compound is not a significant pollution problem. See Fthenakis, Vasilis M., "Life Cycle Impact Analysis of Cadmium in CdTe PV Production," *Renewable and Sustainable Energy Reviews* 8 (2004): 303–334.

CIGS is a chemical compound of copper, indium, gallium, and selenium. Amorphous silicon and microcrystalline S (very small crystals of silicon) are also being used to make thin film photovoltaics, but at the time of this writing, they are more experimental than crystalline silicon or cadmium telluride. An efficiency of 19.9% has been achieved with CIGS, much higher than "Cadmium Telluride (CdTe) or amorphous silicon (a-Si)." (See Wikipedia, http://en.wikipedia.org/wiki/Copper_indium_gallium_selenide; High-efficiency CDTE, accessed 2 June 2009; and

Noufi, Rommel, and Ken Zweibel, *CIGS Thin-film Solar Cells: Highlights and Challenges,* [undated] National Renewable Energy Laboratory Report). But this is still lower than the maximum efficiency of 24% obtained from the more conventional crystalline film silicon oxide photovoltaics (Green, M.A., Jianhua Zhao, A. Wang, and S. R. Wenham, "Very High Efficiency Silicon Solar Cells—Science and Technology," *IEEE Transactions on Electron Devices* 46, no. 10. [1999]: 1940–1947).

4 Lenardic, D., "Large-Scale Photovoltaic Power Plants: Cumulative and Annual Installed Power Output Capacity," 2008. http://pvresources.com.

5 *Ibid.*

6 The Florida Light and Power website provided the information about SEGS. See www.fplenergy.com/portfolio/contents/segs_viii.shtml.

7 *DOE EIA Electric Power Annual,* with data for 2006, 22 October 2007. www.eia.doe.gov/cneaf/electricity/epa/figes1.html.

8 www.solarbuzz.com/StatsMarketshare.htm.

9 Assumptions: Efficiency of solar collectors = 10%; 1 acre = 43,560 sq. ft.; 1 sq. mile = 640 acres; 1 sq. km = 0.3861 sq mile.

10 Lewis, N. S., *Global Energy Perspective,* California Institute of Technology Division of Chemistry and Chemical Engineering PowerPoint presentation http://nsl.caltech.edu/energy.html.

11 Solar Cooker International, Sacramento, CA. See http://solarcookers.org/basics/how.html.

12 Chazan, Guy, "Smaller, Smarter for Remote Areas of Poor Countries, Getting Electricity Doesn't Have to Mean Extending the Grid: There May Be a Simpler Way," *Wall Street Journal,* 11 February 2008, R10.

13 www.epsea.org/pv.html.

14 www.solar2006.org/presentations/tech_sessions/t25-m199.pdf.

15 Hankins, Mark, "A Case Study on Private Provision of Photovoltaic Systems in Kenya," The World Bank. www.worldbank.org/html/fpd/esmap/energy_report2000/ch11.pdf.

16 World Mapper Website, "Households By Nation." www.worldmapper.org/. Accessed 3 January 2010.

17 Chazan, "Smaller, Smarter for Remote Areas of Poor Countries," 2008.

18 www.boeing.com/defense-space/space/bss/hsc_pressreleases/photogallery/spect ro010827/01pr_01494_hirez.jpg.

19 www.solarbuzz.com/SolarPrices.htm.

20 Price to consumers comes from the U.S. Energy Information Administration, "Average Retail Price of Electricity to Ultimate Customers by End-Use Sector, by State," with data for December 2007, 13 March 2008. www.eia.doe.gov/cneaf/electricity/epm/table5_6_a.html.

21 http://en.wikipedia.org/wiki/Solar_power_plants_in_the_Mojave_Desert.

22 Anonymous, "Editorial Wasting and Wanting at the Pentagon," *New York Times,* 2 April 2008.

23 Bilmes, Linda J. and Joseph E. Stiglitz, "The Iraq War Will Cost Us $3 Trillion, and Much More," *Washington Post*, 2 April 2008, B01.

24 Lenardic, Denis, "Photovoltaic Economics," 2008. www.pvresources.com/en/economics.php.

25 Anonymous, *2006–2007 Annual Report on the Development of Global Solar Energy Industry*, 03 June 2007. www.chinaccm.com. Accessed 1 March 2007.

Also see www.chinaccm.com/4S/4S04/4S0401/news/20070301/161316.asp.

26 www.nasa.gov/centers/dryden/news/FactSheets/FS-034-DFRC.html.

Chapter 8

1 Geoghegan, John, "Long Ocean Voyage Set for Vessel That Runs on Wave Power," *New York Times*, 11 March 2008.

2 www.tsuneishi.co.jp/english/horie/about.html.

3 Electric Power Research Institute, "Ocean Thermal Energy Conversion." www.nrel.gov/otec/what.html. Accessed 10 February 2008.

4 World Energy Council, "Survey of Energy Resources 2007: Harnessing the Energy in the Tides." www.oceanpowertechnologies.com/res.htm.

5 Clark, Pete, Rebecca Klossner, and Lauren Kologe. CAUSE 2003 final project. www.ems.psu.edu/~elsworth/courses/cause2003/finalprojects/canutepresentation.pdf. Accessed 9 February 2008. (Quality of information is unknown.)

6 www.tidalelectric.com/History.htm. Accessed 9 February 2008.

7 Jackson, T., and R. Lofstedt, Royal Commission on Environmental Pollution, *Study on Energy and the Environment*. www.rcep.org.uk/ studies/energy/98-6061/jackson.html. Accessed 29 November 2000.

8 Institute of Engineering and Technology, "Tidal Power." http://search.theiet.org/iet/search?action=IETSearch&q=tidal+power

9 Information about La France tidal power plant comes from Chapter 19 of Botkin, D. B., and E. A. Keller, *Environmental Science: Earth as a Living Planet* (New York: John Wiley & Sons, 2009); and from the websites www.reuk.co.uk/La-Rance-Tidal-Power-Plant.htm and www.ems.psu.edu/~elsworth/courses/cause2003/finalprojects/canutepresentation.pdf.

10 Botkin and Keller, *Environmental Science*, 2009.

11 www.ems.psu.edu/~elsworth/courses/cause2003/finalprojects/canutepresentation.pdf.

12 Clery, D., "U.K. Ponders World's Biggest Tidal Power Scheme," *Science* 320 (2008): 1754. This report gives the amount as 17 terawatt-hours of energy per year. See www.sciencemag.org.

13 *Ibid.*

14 Institute of Engineering and Technology, "Tidal Power."

15 Timmons, H., "Energy from the Restless Sea," *New York Times*, 3 August 2006.

16 "The bioSTREAM is a renewable energy technology designed to convert tidal and marine current energy into useful electricity. The power conversion process

and associated device motions are modeled on biological species, such as shark and tuna, that use Thunniform-mode swimming propulsion. By mimicking these creatures, the bioSTREAM benefits from 3.8 billion years of evolutionary hydro-dynamic optimization. The inherited biological traits result in a cost effective and reliable renewable energy system."

17 Timmons, "Energy," 2006.

18 EPRI, "What Is Ocean Thermal Energy Conversion?" www.nrel.gov/otec/what. html; and "Achievements in OTEC Technology," www.nrel.gov/otec/ achievements.html. Accessed 10 February 2008.

19 *Ibid.*

Chapter 9

1 Photos by Debbie Roos, Agricultural Extension Agent, 1 August 2005. www. ces.ncsu.edu/chatham/ag/SustAg/farmphotoaugust0105.html. Co-op member John Bonitz demonstrates how to catch the seeds that shatter during harvest. He is harvesting mustard, one of the many oilseed crops that can be used to create biodiesel fuel.

2 The North Carolina Biodiesel Trade Group was started in 2007 and has its own website. See http://news.biofuels.coop/2008/01/15/north-carolina-biodiesel-trade-group-launched/.

3 ORNL (2008) conversion factors used by the Bioenergy Feedstock Development Programs at ORNL.

4 Jacoby, J., "Sky-High Gas Prices? Not Really," *Boston Globe*, 20 May 2004.

5 World Firewood Supply World Energy Council, www.worldenergy.org. Accessed 24 April 2006.

6 Ndayambaje, J. D., "Agroforestry for Wood Energy Production in Rwanda," Workshop on Alternative Sources of Energy in Rwanda, organized by IRST Cen-tre Iwacu, Kabusunzu, Institut Des Sciences Agronomiques du Rwanda, Recherche forestière et Agroforestière, May 2005.

7 Biran, Adam, Joanne Abbot, and Ruth Mace, "Families and Firewood: A Com-parative Analysis of the Costs and Benefits of Children in Firewood Collection and Use in Two Rural Communities in Sub-Saharan Africa," *Human Ecology* 32, no. 1 (2004): 1–25.

8 Zezima, K., "With Oil Prices Rising, Wood Makes a Comeback," *New York Times,* 19 February 2008. The number of houses using wood for heating comes from census data provided by the DOE EIA.

9 Tuttle, R., "Wood Fuel Pollution Firewood Cost Prices: A View of Things to Come: Environmental Cost of Burning Wood," Bloomberg.com, 2007. www.bloomberg.com/news/.

10 U.S. EPA Office of Air Quality Planning and Standards, Woodstove Changeout Workshop: Nature and Magnitude of the Problem, 8 March 2006.

11 Stix, Gary, "A Climate Repair Manual," *Scientific American* 295 (2006): 46–49.

12 "The Warming Challenge," *New York Times* editorial, 5 May 2007. The editorial stated, "The new report deals with remedies. It warns that over the course of this

century, major investments in new and essentially carbon-free energy sources will be required. But it stresses that we can and must begin to address the problem now, using off-the-shelf technologies to make our cars, buildings, and appliances far more efficient, while investing in alternative fuels, like cellulosic ethanol, that show near-term promise."

13 *Scientific American* online, www.scientificamerican.com/article.cfm?id=jumbo-jet-no-longer-biofuel-virgin-after -palm-oil-flight. Accessed 25 February 2009.

14 www.futurepundit.com/archives/003271.html. Future technological trends and their likely effects on human society, politics, and evolution. (May not be a reliable source of information.)

15 "USDA Biomass Fuels," www.ers.usda.gov/Briefing/Bioenergy/. Accessed 18 July 2007.

16 Barta, P., "Biofuel Costs Hurt Effort to Curb Oil Price," *Wall Street Journal,* 5 November 2007, A2.

17 Mang, H. P., "Biofuel in China," Chinese Academy of Agricultural Engineering (CAAE) Center of Energy and Environmental Protection (CEEP) Ministry of Agriculture PowerPoint presentation, 26 March 2007.

18 Gibbon, G. A., "U.S. Energy Sources and Consumption," www.sc-2.psc.edu/news/USEnergy.ppt. Accessed 10 January 2007.

19 Biomass Gas and Electric Company's website, www.biggreenenergy.com/Default.aspx?tabid=4314. No information about the status of the Port St. Joe project was available as of 25 August 2009.

20 Saslow, L., "From Restaurant Fryers, a Petroleum Alternative," *New York Times,* 4 November 2007.

21 Saulny, Susan, "Greasy Thievery," *New York Times,* 30 May 2008.

22 http://masadaonline.com/.

23 The list of species in use, in test, or proposed comes from the following sources:

(1) Ecological Society of America, "Biofuels: More Than Just Ethanol," *ScienceDaily,* 6 April 2007. www.sciencedaily.com /releases/2007/04/ 070405122400.htm. Accessed 29 January 2008.

(2) South Dakota State University, "Prairie Cordgrass for Cellulosic Ethanol Production," *ScienceDaily,* 28 June 2007. www.sciencedaily.com/releases/2007/ 06/070627122622.htm. Accessed 29 January 2008.

(3) Rosenthal, E., "With Measure of Caution, Europe Joins Biofuel Gold Rush," *New York Times,* 28 May 2007.

24 University of Minnesota, "Fuels Made from Prairie Biomass Reduce Atmospheric Carbon Dioxide," *Science Daily,* 11 December 2006. www.sciencedaily. com /releases/2006/12/061207161136.htm. Accessed 29 January 2008.

25 Etter, L., "With Corn Prices Rising, Pigs Switch to Fatty Snacks on the Menus: Trail Mix, Cheese Curls, Tater Tots; Farmer Jones's Ethanol Fix," *Wall Street Journal,* May 21, 2007.

26 An important discussion of biofuels and food is found in Pimentel, David (ed.), *Biofuels, Solar and Wind Energy as Renewable Energy Systems* (Heidelberg / New York: Springer, 2008).

27 Pimentel, David, Alison Marklein, et al., "Food Versus Biofuels: Environmental and Economic Costs," *Human Ecology* 37 (2009): 1–12.

28 Tilman, D., J. Hill, and C. Lehman, "Carbon-Negative Biofuels from Low-Input High-Diversity Grassland Biomass," *Science* 314 (2006): 1598–1600.

29 Pimentel, D., and T. W. Patzek, "Ethanol Production Using Corn, Switchgrass, and Wood; Biodiesel Production Using Soybean and Sunflower," *Natural Resources Research* 14, no. 1 (2005): 65–76.

30 Hill, J., et al., "Environmental, Economic, and Energetic Costs and Benefits of Biodiesel and Ethanol Biofuels," *PNAS* 103, no. 30 (2006): 11206–11210.

31 Pimentel and Patzek, "Ethanol Production," 2005.

32 Energy Information Administration, Renewable Energy Consumption and Preliminary Statistics 2008.

33 Dale, Bruce E., "Thinking Clearly About Biofuels: Ending the Irrelevant 'Net Energy' Debate and Developing Better Performance Metrics for Alternative Fuels USA," *Biofuels, Bioproducts and Biorefinin* Volume 1, Issue 1, pp. 14-17. Published Online 9 August 2007, www.interscience.wiley.com.

34 U.S. Congress, *Energy Independence and Security Act of 2007* (originally named the *CLEAN Energy Act of 2007*). The Act is titled "An Act to move the United States toward greater energy independence and security; to increase the production of clean renewable fuels; to protect consumers, to increase the efficiency of products, buildings, and vehicles; to promote research on and deploy greenhouse gas capture and storage options; and to improve the energy performance of the Federal Government, and for other purposes." Its sponsor is Rep. Nick J. Rahall II (WV-3). The act requires 144 billion liters of ethanol from biofuels produced each year by 2022.

35 Sinclair, Thomas R., "Taking Measure of Biofuels," *American Scientist* 97, no. 5 (2009): 400–407.

36 USDA press release, "U.S. Crop Acreage Down Slightly in 2009, but Corn and Soybean Acres Up," June 2009. www.nass.usda.gov/Newsroom/2009/06_30_2009.asp. In 2009, American farmers planted 321 million acres.

37 The material on phosphate mining comes from Botkin D. B., and E. A. Keller, *Environmental Science: Earth as a Living Planet* (New York: John Wiley & Sons, 2009).

38 Vaccari, D. A., "Phosphorus: A Looming Crisis," *Scientific American* 300, no. 6 (2009): 54–59.

39 Smil, V., "Phosphorus in the Environment: Natural Flows and Human Interference," *Annual Review of Energy and the Environment* 25 (2000): 53–88.

40 U.S. Geological Survey 2009, http://minerals.usgs.gov/minerals/pubs/commodity/phosphate_rock/mcs-2009-phosp.pdf.

41 *Ibid.*

42 Pimentel, D., personal communication with the author, 19 August 2009.

43 www.sunecoenergy.com/index.cfm?page=pages&pages_ID=9.

44 Johnson, K., "A New Test for Business and Biofuel," *New York Times*, 17 August 2009, A3.

45 Warnick, T. A., and B. A. Methe, et al., "Clostridium Phytofermentans Sp. Nov., a Cellulolytic Mesophile from Forest Soil," *International Journal of Systematic and Evolutionary Microbiology* 52 (2002): 1155–1160.

46 Dias de Oliveira, M. E., Burton E. Vaughan, and Edward J. Rykiel, Jr., "Ethanol as Fuel: Energy, Carbon Dioxide Balances, and Ecological Footprint," *Bio-Science* 55, no. 7 (2005): 593.

47 Boddey, R. M., L. H. de B. Soares, et al., "Bio-Ethanol Production in Brazil: Bio-fuels, Solar and Wind Energy as Renewable Energy Systems," in D. Pimentel (ed.) *Biofuels, Solar and Wind as Renewable Energy Systems: Benefits and Risks,* Heidelburg/New York: Springer: 321–356. These authors state that the net energy return in Brazil from ethanol is 8.8 (8.8 times as much energy is obtained from the resulting fuel than was used to produce it). They note that Pimentel gets a very different value of 1.66. The difference, they state, is due to a great difference in the estimate of the energy to transport fertilizers and chemicals to the cropland and sugar cane to the mills.

48 Dias de Oliveira, Burton, and Rykiel, Jr., "Ethanol as Fuel: Energy," 2005.

49 IPCC, *IPCC Fourth Assessment Report,* report to Intergovernmental Panel on Climate Change, Valencia, Spain, 2007.

50 Kanter, J., "Europe May Ban Imports of Some Biofuel Crops," *New York Times,* 15 January 2008.

51 *Ibid.* (44 million acres is 18 million hectares.)

52 Barta, P., "Biofuel Costs Hurt Effort to Curb Oil Price," *Wall Street Journal,* 5 November 2007, A2.

53 Sumner, Daniel A. and Henrich Brunke, *The Economic Contributions of the California Rice,* 6 May 2006. California Rice Commission at www.calrice.org/c3a_economic_impact.htm.

54 Rosenthal, E., "With Measure of Caution, Europe Joins Biofuel Gold Rush," *New York Times,* 28 May 2007.

55 Barber, J., *Policy Gap Analysis: Findings & Policy Recommendations for the Biomass Sector,* USDA, 2007. The report states, "Biodiesel gets a direct subsidy of $0.50 per gallon, and $1.00 a gallon for "agribiodiesel and renewable diesel." I interpret this to mean 50¢ per gallon for nonagricultural biodiesel and $1 per gallon for biodiesel produced on farms.

56 Reported as $1.24 per liter, while the cost to produce a liter of gasoline from fossil fuels was 33¢ per liter. McCain, 2003, quoted in Pimentel and Patzek (2005), reports that including the direct subsidies for ethanol plus the subsidies for corn grain, a liter costs 79¢ ($3 per gallon). If the production costs of producing a liter of ethanol were added to the tax subsidies, the total cost for a liter of ethanol would be $1.24. Because of the relatively low energy content of ethanol, 1.6 liters of ethanol have the energy equivalent of 1 liter of gasoline. Thus, the cost of producing ethanol to equal a liter of gasoline is $1.88 ($7.12 per gallon of gasoline), while the current cost of producing a liter of gasoline is 33¢ (USBC, 2003).

57 According to Pimentel and Patzek (2005), government subsidies "total more than 79¢/l [cents per liter, which is $3 per gallon] are mainly paid to large corporations (McCain, 2003). To date, a conservative calculation suggests that corn farmers are receiving a maximum of only an added 2¢ per bushel for their corn or less than

$2.80 per acre because of the corn ethanol production system. Some politicians have the mistaken belief that ethanol production provides large benefits for farmers, but in fact the farmer profits are minimal."

58 McCain 2003, in Pimentel and Patzek, "Ethanol Production," 2005.

59 National Center for Policy Analysis, in Pimentel and Patzek, 2005. About 70% of the corn grain is fed to U.S. livestock (USDA, 2003a, 2003b).

60 Pimentel and Patzek, "Ethanol Production," 2005.

61 Botkin, D. B., and Charles R. Malone, "Efficiency of Net Primary Production Based on Light Intercepted During the Growing Season," *Ecology* 40 (1968): 439–444. About 3% stored by the old field's vegetation is a lot better than desert ecosystems, whose vegetation is able to store less than 0.03% of the sunlight, but similar to what has been found for various forests.

62 University of Minnesota, "Fuels Made from Prairie Biomass Reduce Atmospheric Carbon Dioxide," *ScienceDaily* (11 December 2006). www.sciencedaily.com/releases/2006/12/061207161136.htm. Accessed 29 January 2008.

63 Mang, H. P., *Biofuel in China*. Chinese Academy of Agricultural Engineering (CAAE), Center of Energy and Environmental Protection (CEEP), Ministry of Agriculture PowerPoint presentation (March 2007).

64 *Ibid.*

65 http://news.biofuels.coop/2008/01/15/north-carolina-biodiesel-trade-group-launched/.

66 Burton, Rachel and Leif Forer, "Introduction to Biofuels: Biodiesel and Straight Vegetable Oil," Biofuels Program, Central Carolina Community College, Pittsboro, NC, 2007.

67 Dias de Oliveira, M. E., Burton E. Vaughan, and Edward J. Rykiel, Jr., "Ethanol as Fuel: Energy, Carbon Dioxide Balances, and Ecological Footprint," *BioScience* 55, no. 7 (2005): 593.

Chapter 10

1 The gasoline pipeline explosion stories come from Cat Lazaroff, "Negligence Caused Pipeline Explosion, Suit Charges," Environmental News Service, 31 May 2002.

2 Trench, C. J., "How Pipelines Make the Oil Market Work—Their Networks, Operation, and Regulation," memorandum prepared for the Association of Oil Pipe Lines and the American Petroleum Institute's Pipeline Committee, New York, 2001.

3 See Lazaroff, "Negligence Caused Pipeline Explosion, Suit Charges," 2002.

4 www.pipeline101.com/Overview/crude-pl.html.

5 Trench, "How Pipelines Make the Oil Market Work—Their Networks, Operation, and Regulation," 2001.

6 www.pipeline101.com/Overview/crude-pl.html.

7 Estimated from the Association of Oil Pipe Lines, *Shifts in Petroleum Transportation,* 2000.

8 R. A. Wilson, *Transportation in America,* 18th edition (Washington, D.C.: Eno Transportation Foundation, Inc., 2001). "How Pipelines Make the Oil Market Work: Pipelines are Key to Meeting U.S. Oil Demand Requirements Allegro Energy Group."

9 Trench, "How Pipelines Make the Oil Market Work—Their Networks, Operation, and Regulation," 2001.

10 North American Electric Reliability Corporation, *Long-Term Reliability Assessment 2007—2016.* (Princeton, NJ: North American Electric Reliability Corporation, 2007).

11 *Ibid.*

12 American Gas Association, www.aga.org/Kc/aboutnaturalgas/consumerinfo/NGDeliverySystemFacts.htm.

13 EIA, www.eia.doe.gov/pub/oil_gas/natural_gas/analysis_publications/ngpipeline/. The U.S. natural gas pipeline network is a highly integrated transmission and distribution grid that can transport natural gas to and from nearly any location in the lower 48 states. The natural gas pipeline grid comprises more than 210 natural gas pipeline systems, 302,000 miles of interstate and intrastate transmission pipelines, and more than 1,400 compressor stations that maintain pressure on the natural gas pipeline network.

14 DOE EIA, www.eia.doe.gov/pub/oil_gas/natural_gas/analysis_publications/ngpipeline/ngpipe lines_map.html.

15 Trench, "How Pipelines Make the Oil Market Work—Their Networks, Operation, and Regulation," 2001.

16 Anderson, Roger, and Albert Boulanger, "Smart Grids and the American Way," *Mechanical Engineering Power & Energy* (March 2004). Online at http://www.memagazine.org/supparch/pemar04/smgrids/smgrids.html.

17 Tverberg, Gail E., "The U. S. Electric Grid: Will It Be Our Undoing?" The Oil Drum, 7 May 2008. www.energybulletin.net/node/43823. This report states that most energy transformers on the grid are more than 40 years old.

18 Owens, D. K., "Electricity: 30 Years of Industry Change, 30 Years of Energy Information and Analysis," 7 April 2008. Edison Electric Institute, The Association of Shareholder-Owned Electric Companies.

19 North American Electric Reliability Corporation, 2007.

20 Gridpoint website, www.gridpoint.com/news/press/20080514.aspx.

21 Anderson and Boulanger, "Smart Grids and the American Way," 2004.

22 Gelsi, Steve, "Power Firms Grasp New Tech for Aging Grid," *MarketWatch,* 11 July 2008.

23 Dunn, S., *Hydrogen Futures: Toward a Sustainable Energy System,* Worldwatch paper 157, 2001.

24 World Wind Energy, www.world-wind-energy.info/.

25 Rifkin, J., *The Hydrogen Economy* (New York: Tarcher, 2003).

Chapter 11

1 Chmielewski, Dawn C., and Ken Bensinger, "Automakers Are Lining Up Celebrities to Promote the Technology, but the Clean Fuel Isn't Ready for Prime Time," *L. A. Times*, 15 June 2008.

2 ©2008 Business Wire.

3 www.dot.gov/affairs/dot8408.htm.

4 EIA kids' energy page, www.eia.doe.gov/kids/energyfacts/uses/transportation. html.

5 Amtrak consumed 14.6 trillion BTUs, which is 4.28 billion kilowatt-hours, in 2005. (See www.narprail.org/cms/index.php/resources/more/oak_ridge_fuel/.) Total U.S. freight transportation is 4.23 trillion ton-miles (2005 value), excluding gas and liquids transported by pipelines. At 0.1 kilowatt-hours per ton-mile, this requires a total energy of 423 billion kilowatt-hours, which is only 5.1% of the total energy used in transportation. This doesn't seem right. How could passenger travel use up 94.9%? Try this another way: Transportation uses 28% of the total energy used in the United States, which is 8,203 billion kilowatt-hours. Of the total energy used in transportation, 21% is used to transport coal—1,821 billion kilowatt-hours.

6 One gallon of gasoline contains 29 or 33 kilowatt-hours of energy (depending on whose information you believe). A gallon of diesel fuel contains 40.6 kilowatt-hours. According to Oak Ridge, a gallon of diesel fuel contains 138,700BTUs. One BTU is 2.93 × 10-4 kilowatt-hours. This gives the 40.6 kilowatt-hours per gallon of diesel fuel.

7 www.marketwatch.com/news/story/americans-drive-11-billion-fewer/story. aspx?gu id=%7B93E83ED2-0EE6-48BF-B104-D82FE8A93D70%7D&dist= msr_9.

8 According to the DOT 2002 economic report, a railroad train can carry a ton ten miles for 1 kilowatt-hour, and the coal carried by train in 2002 totaled 590 billion ton-miles, which would have required 59 billion kilowatt-hours.

9 The data comes from U.S. Department of Transportation, *Summary of Fuel Economy Performance*, March 2004. www.dailyfueleconomytip.com/ miscellaneous/average-gas-mileage-relatively-flat-between-1980-and-2004/. According to the National Highway Traffic Safety Administration (NHTSA), the average gas mileage for new vehicles sold in the United States went from 23.1 miles per gallon (mpg) in 1980 to 24.7 mpg in 2004. This represents a paltry increase of slightly less than 7% over the 25-year period.

10 U.S. Congress, *Energy Independence and Security Act of 2007*. "The Secretary shall prescribe a separate average fuel economy standard for passenger automobiles and a separate average fuel economy standard for nonpassenger automobiles for each model year beginning with model year 2011 to achieve a

combined fuel economy average for model year 2020 of at least 35 miles per gallon for the total fleet of passenger and non-passenger automobiles manufactured
for sale in the United States for that model year."

11 According to the Department of Transportation, "Americans drove 1.4 billion
fewer highway miles in April 2008 than in April 2007. While fuel prices and transit ridership are both on the rise, sixth month of declining vehicle miles traveled
signals need to find new revenue sources for highway and transit programs,
Transportation Secretary Mary E. Peters says" 18 June 2008. www.dot.gov/
affairs/dot8408.htm.

12 American Society of Civil Engineers, "Report Card for America's Infrastructure,"
2005.

13 Tom Payne, railroad executive and expert, personnel communication with the
author, June, 2008.

14 According to an 1881 *New York Times* article, in that year, construction of railroads cost $25,000 a mile and 9,358 miles were built, at a total cost of
$233,750,000. Translated into today's dollars (figuring an average inflation of 5%
per year), this same construction would cost $119 billion in 2008, or $12.7 per
mile. Today a plan to install a French *grande vitesse* rail line from Houston to
Dallas—250 miles—was going to cost $22 million a mile, or a total of $5.7 billion.
Amtrak estimates that a new high-speed rail line from Connecticut to Rhode
Island would cost $36 million a mile.

15 www.colorado.edu/libraries/govpubs/dia.htm.

16 American Society of Civil Engineers, 2005.

17 *Ibid.*

18 U.S. Department of Transportation Fiscal Year 2011 Budget Highlights, published 1 February 2010. www.dot.gov/budget/2010/2011budgethighlights.pdf.

19 U.S. Department of Transportation, Bureau of Transportation Statistics, National
Transportation Statistics. www.bts.gov/publications/national_transportation_
statistics/.

20 *Ibid.*

21 The Acela gets 0.833 passenger miles per BTU.

22 www.minnesotarailroads.com/news.html leads to a PowerPoint presentation by
Minnesota Railroads that contains this statement that railroads use about one-
third as much energy per mile as trucks.

Chapter 12

1 DOE, "Building America Habitat Metro Denver," 29 September 2008.
www.eere.energy.gov/buildings/building_america/pdfs/36102.pdf.

2 Courtesy of DOE/NREL. Photo by Pete Beverly. www.nrel.gov/data/pix/Jpegs/
14163.jpg.

3 This image has been reprinted from the National Renewable Energy Laboratory. "Zero Energy Homes Research: A Modest Zero Energy Home," 2009. www.nrel.gov/buildings/zero_energy.html. Accessed December 29, 2009.

4 Geiger, R., *The Climate Near the Ground* (Cambridge, Mass.: Harvard University Press, 1950).

5 Gates, D. M., *Energy Exchange in the Biosphere* (New York, Harper & Row, 1962).

6 DOE, "Building America Habitat Metro Denver," 2008.

7 Drake-McDonough, C., "Home, Sweet (Green) Home: Developments in Three Communities Work to Attract Buyers by Being Kind to the Environment," *Denver Post,* 21 September 2008.

8 Based on the study Katz, Greg, "The Cost and Financial Benefits of Green Buildings, 2003."

9 Another NREL analysis can be found by Anderson, R., C. Christensen, and S. Horowitz, "Program Design Analysis using BEopt Building Energy Optimization," Conference paper presented at the 2006 ACEEE Summer Study on Energy Efficiency in Buildings. http://www.google.com/webhp?rls=ig#hl= en&rls=ig&rlz=1R2SNNT_enUS354&q=Program+Design+Analysis+using+ BEopt+Building+Energy+Optimization&aq=&aqi=&oq=Program+Design+ Analysis+using+BEopt+Building+Energy+Optimization&fp=baa94940edcea411

10 Drake-McDonough, "Home, Sweet (Green) Home," 2008.

11 Howard, E., *Garden Cities of Tomorrow* (reprint). Cambridge, Mass.: MIT Press, 1965.

12 Botkin, D. B., and C. E. Beveridge, "Cities as Environments," *Urban Ecosystems* 1 no. 1 (1997): 3–20.

13 *Ibid.*

14 *Ibid.*

15 *Ibid.*

16 McHarg, I. L., *Design with Nature* (New York: John Wiley & Sons: 1969).

17 You can see the current list of AIA green building awards at www.aiatopten. org/hpb/.

18 The 2008 green building condominium winner was Macallen Building Condominiums (Burt Hill with Office dA), in Boston.

19 Renewable Energy World, "Washington State Law Mandates Green Building," 21 April 2005. www.renewableenergyworld.com/rea/news/story?id=25765.

20 Fossil fuels plus nuclear energy provide 94% of the energy used in the United States.

21 Green, B. D., and R. Gerald Nix, *Geothermal—The Energy Under Our Feet: Geothermal Resource Estimates for the United States,* N. R. E. Laboratory, 2006.

22 For example, see the Go Green Development Consortium, Inc., at www.corbinenterprises.com.

23 www.charlies-web.com/genealogy3/txtx541.html. Accessed 12 August 2005.

24 This image has been reprinted from Green and Nix, *Geothermal*, 2006.

25 According to Doug Rye, Licensed Architect, and Phillip Rye, Licensed Civil Engineer, of Doug Rye and Associates. See www.dougrye.com/geothermal.html.

26 Green and Nix, *Geothermal*, 2006.

27 *Ibid.*

28 *Ibid.*

29 www.eia.doe.gov/emeu/aer/pdf/pages/sec2_2.pdf.

30 This image has been reprinted from Green and Nix, *Geothermal*, 2006.

Chapter 13

1 U.S. Census Bureau, Table 1: "Projections of the Population and Components of Change for the United States: 2010 to 2050 (NP2008-T1)," 14 August 2008. www.census.gov/population/www/projections/summarytables.html.

2 Total energy use would increase by 11,718 billion kilowatt-hours, from 29,297 billion kilowatt-hours to 41,016, or, rounding off, to 40 trillion kilowatt-hours from 30 billion kilowatt-hours.

3 Add to this that hydropower would have to increase by 27% to 1,177 from 861 billion kilowatt-hours, which is unlikely.

4 North American Electric Reliability Corporation, *Long-Term Reliability Assessment 2007—2016* (Princeton, NJ: North American Electric Reliability Corporation, 2007).

5 Energy Information Administration, *Annual Energy Outlook 2001*, (Washington, D.C.: December 2000).

6 Then we use the average annual kilowatt-hour output from an installed kilowatt capacity, which, as I previously discussed, is 2.347 kilowatt-hour over a year from an installed watt of wind turbine, and 1.245 kilowatt-hour per installed watt of solar photovoltaics.

7 To calculate the required installed capacity for Scenario 2, you take the ratio of the desired energy output divided by the yield. Yield is the kilowatt-hours produced annually per installed watt:

For wind: I = 15,215/2,347 = 6.48 billion KW

For solar: I = 15,214/1,235 = 12.3 billion KW

Additional Tables: Table 13: Power and Energy Costs of Coal, Solar, and Wind

Power and Energy costs	Daniel B. Botkin	10-Jun-09		
Cost ($) per watt-hour	Cents/Wh/W	$ per kWh		Duration (yrs)
Wind @ $1/w and yields 2347Wh/W over a year				
$ 0.00043	0.042607584	$	0.43	1
		$	0.04	10
		$	0.021	20
		$	0.00043	100
Wind @ $1.17/w and yields 2347Wh/W over a year				
$ 0.00073	0.072858969	$	0.73	1
		$	0.07	10
		$	0.036	20
		$	0.00073	100
Solar @ $1/w and yields 1235 Wh/W over a year				
$ 0.000810	0.08097166	$	0.81	1
		$	0.08	10
		$	0.040	20
		$	0.0081	100
Solar @ $3/w and yields 1235 Wh/W over a year				
$ 0.002429	0.24291498	$	2.43	1
		$	0.24	10
		$	0.12	20
		$	0.024	100
Solar @ $6.81/w and yields 1235 Wh/W over a year				
$ 0.005514	0.551417004	$	5.51	1
		$	0.55	10
		$	0.28	20
		$	0.055	100
Coal @ $10.50/ton and 2000 kWh/ton				
2000 kilowatt hours per ton				
Coal would have to cost B per ton to equal Solar @ $3/W installed over 10 years	$ 462.70			
Coal would have to cost B per ton to equal Solar @ $3/W installed over 20 years	$ 46.27			
Coal would have to cost B per ton to equal Wind @$1/W over 10 years	$ 8.12			
Coal would have to cost B per ton to equal Wind @$1/W over 20 years	$ 4.06			
Coal would have to cost B per ton to equal Solar @ $6.81/W installed over 10 years	$ 105.03			
Coal would have to cost B per ton to equal Solar @ $6.81/W installed over 20 years	$ 52.52			
Coal would have to cost B per ton to equal Wind @$1.71/W over 10 years	$ 13.88			
Coal would have to cost B per ton to equal Wind @$1.71/W over 20 years	$ 6.94			

8 Professor Matthew J. Sobel, William E. Umstattd, professor of operations research at Case Western Reserve University, kindly provided the economic cost analysis, using a social discount factor, for the scenarios in this chapter. He also provided this brief introduction to the concept (25 August 2009):

Alternative energy policies would induce alternative time streams of costs. What is a "fair" comparison of these time streams? Here, I estimate the annual costs (in 2009 dollars) of each energy policy between 2010 to 2050.

Some of the time streams are very simple because they have the same dollar amount every year. For example, the series of costs labeled "Wind Component 3" has the entry $80,500,000,000 each year from 2010 to 2050. Similarly, in "Solar Component 3" the cost is $888,750,000,000 each year from 2010 to 2050. If all the policies induced time streams like these two, then the comparison of the time streams would reduce to the comparison of the constant annual entries. So the comparison of the costs of Wind Component 3 with Solar Component 3 would reduce to the comparison of $80,500,000,000 with $888,750,000,000. We would conclude that the cost of the latter is approximately eleven times the cost of the former.

However, the series of costs labeled "Coal Replaces Fossil Fuels 3" does not have the same cost every year. The cost is $573,029,861,670 in 2010, $582,267,374,847 in 2011, and so on. What is a "fair" comparison of this series with the series of costs of other energy policies? Most economists would say that the simplest useful answer to this question is "net present value" or NPV for short.

The arithmetic of NPV is the same as that of compound interest. That is, if you have a savings account that pays 5% interest each year, then (ignoring taxes) you would have to deposit $1/(1.05) now in order to have $1 one year from now, you would have to deposit $1/(1.05)2 to have $1 two years from now, and you would have to deposit $1/(1.05)2 to have $1 t years from now.

When analyzing public policies, the interest rate (5% in the previous example) is called the social discount rate. Its determination can be a matter of dispute in particular situations, but it is often between 3% and 7% in developed countries like the U.S. So I use 5% here. The NPV of the series of costs labeled "Wind Component 3" is [$80,500,000,000 × 1] + [$80,500,000,000 / 1.05] + [$80,500,000,000/(1.05)^2 + [$80,500,000,000/(1.05)^3 + ...+ [$80,500,000,000 / (1.05)^{40}]. The first bracketed amount is for the year 2010, the second is for 2011, the third for 2012, the fourth for 2013, and the last for 2050. The NPV is the sum $1,461,806,451,497.

Similarly, the NPV of "Coal Replaces Fossil Fuels 3" is [$573,029,861,670] + [$582,267,374,847/1.05] + [$591,504,888,024 / (1.05)^2 + ... + [$942530,388,742 / (1.05)^{40} = $12,684,632,664,159. So a "fair" comparison of the series of costs of "Coal Replaces Fossil Fuels 3" with those of "Wind Component 3" is that the NPV of the former is approximately 8.7 times the NPV of the latter.

9   The 2009 federal budget summary document is titled *A New Era of Responsibility: Renewing America's Promise,* and is available at www.gpoaccess.gov/usbudget/fy10/pdf/fy10-newera.pdf. Accessed 29 June 2009.

10 *Ibid.*

11 UNEP, *Global Trends in Sustainable Energy Investment 2009: Analysis of Trends and Issues in the Financing of Renewable Energy and Energy Efficiency.*

12 Mouawad, Jad, and Kate Galbraith, "By Degrees: Plugged-In Age Feeds a Hunger for Electricity," *New York Times,* 20 September 2009.

13 *Ibid.*

14 Solar installations produce 1,235 kilowatt-hours per year for each kilowatt installed; wind generates 2,347 kilowatt-hours per year for each kilowatt installed.

From this, we can calculate the installation required as given in the text. Again, we use the average annual kilowatt-hour output from an installed kilowatt capacity, which are 2,347 watt-hours or 2.347 kilowatt-hours over a year from an installed watt of wind turbine, and 1,235 watt-hours or 1.245 kilowatt-hours per installed watt of solar photovoltaics. Based on these generation amounts, in 2050, solar capacity would have to be 5.22 billion kilowatts and wind energy capacity would have to be 2.75 billion kilowatts for each to produce 6,448 billion kilowatt-hours, the energy requirements for these sources in Scenario 3.

15 Solar installations produce 1,235 kilowatt-hours per year for each kilowatt installed; wind generates 2,347 kilowatt-hours per year for each kilowatt installed. From this, we can calculate the installation required as given in the text. At a cost of $6,810 per kilowatt, solar will cost $5.22 \times 10 \times \$6,810 = \$35.55$ trillion, which, spread over 40 years, is $899 billion a year. At an installation cost of $1,170 per kilowatt for wind, the wind energy installations would cost $5.76 trillion, or $144 billion per year. Combined, solar and wind capacity would cost $1,033 billion a year. Note that this assumes that wind and solar share equally in replacing most fossil fuels.

16 EIA, Table 2.17: "Annual Photovoltaic Domestic Shipments, 1997–2006." www.eia.doe.gov/cneaf/solar.renewables/page/solarphotv/solarpv.html.

17 According to the Electric Power Research Institute (EPRI), a 100MW wind energy farm would require 25,333 acres, or about 40 square miles. Therefore, 1MW requires 0.4 sq. miles and Scenario 3's 2.88 billion kilowatts requires 1152 sq. miles. However, actual installations vary. For example, the two largest wind farms in Texas have a very different turbine density and, therefore, different energy capacity density. Roscoe Wind Farm has a stated output capacity 6% greater than Horse Hollow's, but it takes up more than twice the area. Both are more widely spread out than the average estimate from EPRI. At Roscoe's density, for wind turbines to provide the electric power for all homes in the United States, it would require 2.6% of the lower 48's land area; at Horse Hollow's, 1.3%.

18 Urban area of lower 48 comes from Economic Research Service, *Major Uses of Land in the United States, 2002*. www.ers.usda.gov/publications/EIB14/eib14g.pdf. Cropland data comes from www.ers.usda.gov/Data/MajorLandUses/.

19 Calculations for Scenario 3. Without the social discount factor: For wind, the installed capacity would be 2.75 billion kilowatts. For solar, it would be 5.22 billion kilowatts. At a cost of $1,170 per kilowatt, wind will cost $5.76 trillion. At a cost of $6,810 per kilowatt, solar will cost $35.55 trillion. These are corrected in the text to take into account an annual 5% social discount factor, as discussed.

20 Smith, R., "U.S. Foresees a Thinner Cushion of Coal," *Wall Street Journal*, 8 June 2009.

21 Oster, Shai, and Ann Davis, "China Spurs Coal-Price Surge Once-Huge Exporter Now Drains Supply; Repeat of Oil's Rise?" *Wall Street Journal*, 12 February 2008., A1.

22 *Ibid.*

23 Most recent cost estimates to build nuclear power plants come from Michael Totty, 2008. "The Case For and Against Nuclear Power WSF," *Wall Street Journal*, 20 June 2008.

24 The 2008 cost of uranium ore averaged $45.88 a pound, and 53 million pounds were purchased by civilian nuclear power plants, costing a total of $2.3 billion a year. If we wanted to go the nuclear route for Scenario 3, the fuel required in 2050 would cost, in current dollars, $13.4 billion a year.

25 The value of the fuel to provide the present energy output from the 104 nuclear reactors in the United States, including mining, refining, and all other production costs, is $1.17 trillion, based on the cost of 45¢ per kilowatt-hour, given by the World Nuclear Association.

26 Information on decommissioning nuclear power plants comes from the U.S. Nuclear Regulatory Commission, "Fact Sheet on Decommissioning Nuclear Power Plants." www.nrc.gov/reading-rm/doc-collections/fact-sheets/decommissioning.html. See also www.nrc.gov/info-finder/decommissioning/power-reactor/.

27 Wald, Matthew J., "Dismantling Nuclear Reactors," *Scientific American*, 26 January 2009. www.scientificamerican.com/article.cfm?id=dismantling-nuclear.

28 World Nuclear Association.

29 Alexander, Lamar, Chairman, "Blueprint for 100 New Nuclear Power Plants in 20 Years" (Washington, D.C.; U. S. Senate Republican Conference, 2009).

30 The cost comparison for the variations of Scenario 3 are given in this table:

Summary Table of Costs: Adding Coal Pollution Costs ($ Trillions)

	Installation	Fuel for 40 Years	Pollution Costs	Total	Total/ Year
Wind & Solar Provide 65%	$48.8			$48.8	$1.220
Wind Provides All Energy	$13.1			$13.1	$0.328
Coal Provides 65%	$3.9	$2.9	$28.0	$34.8	$0.870
Nuclear Replaces Fossil Fuels, Adding 900 New Plants	$12.6	$3.4		$16.0	$0.400

Note that, in this table, the cost of a nuclear power plant is assumed to go up and be at the more expensive end of the range, $14 billion each.

31 EIA, www.eia.doe.gov/emeu/aer/consump.html.

32 For more specifics about the terms related to the discussion here, refer to the EIA's online glossary, www.eia.doe.gov/glossary/glossary_i.htm.

33 Methane is CH OH.

34 Lewis, Nathan S., California Institute of Technology Division of Chemistry and Chemical Engineering Pasadena. See http://nsl.caltech.edu.

35 U.S. Congress, *Energy Independence and Security Act of 2007.* "The Secretary shall prescribe a separate average fuel economy standard for passenger automobiles and a separate average fuel economy standard for nonpassenger automobiles for each model year beginning with model year 2011 to achieve a combined fuel economy average for model year 2020 of at least 35 miles per gallon for the total fleet of passenger and non-passenger automobiles manufactured for sale in the United States for that model year."

36 American Society of Civil Engineers, *Report Card for America's Infrastructure,* 2005. www.asce.org/reportcard/2005/index2005.cfm.

37 Taylor, N., *American-Made: The Enduring Legacy of the WPA* (New York: Bantam Books, 2008).

38 Herbert Hoover Library, http://hoover.archives.gov/exhibits/Hooverstory/gallery04/gallery04.html.

39 Taylor, *American-Made,* 2008, p. 55.

40 Taylor, *American-Made,* 2008, p. 107.

Index

A

AASHTO (American Association of State Highway and Transportation Officials), 217
active energy-saving features, 231
agrifuels, 180-181, 273
AID (Agency for International Development), 156
air pollution from burning coal as fuel, 62-65
air travel, 159, 220
Alexander, Lamar, 137, 262
algae, producing biofuel from, 186-188
Allegheny Company, 3, 7
Alliance for Mindanao Offgrid Renewable Energy (AMORE), 156
Alliance for Nuclear Responsibility, 100
Altamont Pass Wind Farm (CA), 132
American Association of State Highway and Transportation Officials (AASHTO), 217

American Wind Energy Association
  estimate of U.S. wind energy potential, 124
  on market for small wind turbines, 135
American-Made: The Enduring Legacy of the WP (Taylor), 267
Anasazi dwellings, 227
ancient Greece/Rome, energy use in, 11-12
Anderson, Roger, 206
Anthracite coal, 54
AMORE (Alliance for Mindanao Offgrid Renewable Energy), 156
Aramco, agreement with Dow Chemical, 33
architecture, energy-efficiency design in
  active methods, 231
  climate/ecology and energy use, 232-233
  costs of, 234-235
  explained, 228-230
  geothermal energy, 239-243
  green buildings, 237-239

X-Y-Z